ecture Notes in Mathematics

:ollection of informal reports and seminars
:ted by A. Dold, Heidelberg and B. Eckmann, Zürich

:ies: Forschungsinstitut für Mathematik, ETH, Züricl

4

Goro Shimura

Princeton University, Princeton, New Jersey

Automorphic Functions and Number Theory

968

nger-Verlag Berlin · Heidelberg · New York

Library of Congress Catalog Card Number 68-25132. Printed in Germany. Title No. 7374

Preface

These notes are based on lectures which I gave at the Forschungsinstitut für Mathematik, Eidgenössische Technische Hochschule, Zürich in July 1967. I have attempted to make a short comprehensible account of the latest results in the field, together with an exposition of the material of an elementary nature. No detailed proofs are given, but there is an indication of basic ideas involved. Occasionally even the definition of fundamental concepts may be somewhat vague. I hope that this procedure will not bother the reader. Some references are collected in the final section in order to overcome these shortcomings. The reader will be able to find in them a more complete presentation of the results given here, with the exception of some results of §10, which I intend to discuss in detail in a future publication.

It is my pleasure to express my thanks to Professors K. Chandrasekharan and B. Eckmann for their interest in this work, and for inviting me to publish it in the Springer Lecture Notes in Mathematics. I wish also acknowledge the support of the Eidgenössische Technische Hochschule, Institute for Advanced Study, and the National Science Foundation (NSF-GP 7444, 5803) during the summer and fall of 1967.

Princeton, January 1968 G. Shimura

Preface

[illegible] notes [illegible] which [illegible]

[illegible] methods for [illegible], [illegible] the

[illegible] (196[illegible]) [illegible] have attempted [illegible]

[illegible] complement [illegible] in the field [illegible]

[illegible] an exposition of the material of an elementary nature [illegible]

[illegible] results are given, but [illegible] an indication of basic ideas

involved. [illegible] the definition of fundamental concepts

[illegible] procedure [illegible]

[illegible] references [illegible] concerned with [illegible] section in

[illegible] The reader will [illegible] able to

find [illegible] complete presentation of the results given here

with [illegible] which [illegible] discuss

[illegible]

[illegible] my pleasure [illegible] Professor [illegible]

[illegible] E. [illegible] help [illegible] this work,

and for [illegible] to publish [illegible] Springer [illegible] Notes in

[illegible]

[illegible] advanced [illegible]

[illegible]

Contents

Notation

We denote by Z, Q, R and C respectively the ring of rational integers, the rational number field, the real number field and the complex number field. For an associative ring Y with identity element, $Y^\times$ denotes the group of invertible elements in Y, $M_n(Y)$ the ring of all matrices of size n with entries in Y, and $GL_n(Y)$ the group of invertible elements in $M_n(Y)$, i.e., $M_n(Y)^\times$. The identity element of $M_n(Y)$ is denoted by 1_n, and the transpose of an element A of $M_n(Y)$ by ${}^tA$ as usual. When Y is commutative, $SL_n(Y)$ denotes the group of all elements of $M_n(Y)$ of determinant 1. For a typographical reason, the quotient of a space S by a group G will be denoted by S/G, even if G acts on the left of S. If F is a field and x is a point in an affine (resp. a projective) space, then $F(x)$ means the field generated over F by the coordinates (resp. the quotients of the homogeneous coordinates) of x. If K is a Galois extension of F, $G(K/F)$ stands for the Galois group of K over F.

1. Introduction

Our starting point is the following theorem which was stated by Kronecker and proved by Weber:

Theorem 1. *Every finite abelian extension of* Q *is contained in a cyclotomic field* $Q(\zeta)$ *with an* m*-th root of unity* $\zeta = e^{2\pi i/m}$ *for some positive integer* m.

As is immediately observed, ζ is the special value of the exponential function $e^{2\pi iz}$ at $z = 1/m$. One can naturally ask the following question:

Find analytic functions which play a role analogous to $e^{2\pi iz}$ *for a given algebraic number field.*

Such a question was raised by Kronecker and later taken up by Hilbert as the 12th of his famous mathematical problems. For an imaginary quadratic field K, this was settled by the works of Kronecker himself, Weber, Takagi, and Hasse. It turns out that the maximal abelian extension of K is generated over K by the special values of certain elliptic functions and elliptic modular functions. A primary purpose of these lectures is to indicate briefly how this result can be generalized for the number fields of higher degree, making thereby an introduction to the theory of automorphic functions and abelian varieties. I will also include some results concerning the zeta function of an algebraic curve in the sense of Hasse and Weil, since this subject is closely connected with the above question. Further, it should be pointed out that the automorphic functions are meaningful as a means of generating not only abelian but also non-abelian algebraic extensions of a number field. Some ideas in this direction will be explained in the last part of the lectures.

2. Automorphic functions on the upper half plane, especially modular functions

Let $\mathfrak{H}$ denote the complex upper half plane:

$$\mathfrak{H} = \{ z \in C \mid \mathrm{Im}(z) > 0 \} .$$

We let every element $\alpha = \begin{pmatrix} a & b \\ c & d \end{pmatrix}$ of $GL_2(R)$, with $\det(\alpha) > 0$, act on $\mathfrak{H}$ by

$$\alpha(z) = (az + b)/(cz + d). \tag{2.1}$$

It is well known that the group of analytic automorphisms of $\mathfrak{H}$ is isomorphic to $SL_2(R)/\{\pm 1_2\}$. Let Γ be a discrete subgroup of $SL_2(R)$. Then the quotient $\mathfrak{H}/\Gamma$ has a structure of Riemann surface such that the natural projection $\mathfrak{H} \rightarrow \mathfrak{H}/\Gamma$ is holomorphic. If $\mathfrak{H}/\Gamma$ is compact, one can simply define an <u>automorphic function on $\mathfrak{H}$ with respect to</u> Γ to be a meromorphic function on $\mathfrak{H}$ invariant under the elements of Γ. Such a function may be regarded as a meromorphic function on the Riemann surface $\mathfrak{H}/\Gamma$ in an obvious way, and vice versa. We shall later discuss special values of automorphic functions with respect to Γ for an arithmetically defined Γ with compact $\mathfrak{H}/\Gamma$. But we first consider the most classical group $\Gamma = SL_2(\mathbf{Z})$. Since $\mathfrak{H}/\Gamma$ is not compact in this case, one has to impose a certain condition on automorphic functions. It is well known that every point of $\mathfrak{H}$ can be transformed by an element of $\Gamma = SL_2(\mathbf{Z})$ into the region

$$F = \{z \in \mathfrak{H} \mid |z| \geqq 1, \ |\mathrm{Re}(z)| \leqq 1/2\}.$$

No two distinct inner points of F can be transformed to each other by an element of Γ. Now $\mathfrak{H}/\Gamma$ can be compactified by adjoining a point at infinity. By taking $e^{2\pi iz}$ as a local parameter around this point, we see that $\mathfrak{H}/\Gamma$ becomes a compact Riemann surface of genus 0. Thus we define an automorphic function with respect to Γ to be a meromorphic function on this Riemann surface, considered as a function on $\mathfrak{H}$. In other words, let f be a Γ-invariant meromorphic function on $\mathfrak{H}$. For $\gamma = \begin{pmatrix} 1 & 1 \\ 0 & 1 \end{pmatrix}$, we have $\gamma(z) = z + 1$. Since $f(\gamma(z)) = f(z)$, we can express $f(z)$ in the form $f(z) = \sum_{n=-\infty}^{\infty} c_n e^{2\pi inz}$ with $c_n \in \mathbf{C}$. Now an automorphic function with respect to Γ is an f such that $c_n = 0$ for all $n < n_0$ for some n_0, i.e., meromorphic in the local parameter $q = e^{2\pi iz}$ at $q = 0$. Such a function is usually called a modular function of level one. Since $\mathfrak{H}/\Gamma$ is of genus 0, all modular functions of level one form a rational function field over $\mathbf{C}$. As a generator of this field, one can choose a function j such that

(2.2) $$j(\sqrt{-1}) = 1, \; j((-1 + \sqrt{-3})/2) = 0, \; j(\infty) = \infty.$$

Obviously the function j can be characterized by (2.2) and the property of being a generator of the field of all modular functions of level one.

Now let K be an imaginary quadratic field, and $\mathfrak{a}$ a fractional ideal in K. Take a basis $\{\omega_1, \omega_2\}$ of $\mathfrak{a}$ over Z. Since K is imaginary, ω_1/ω_2 is not real. Therefore one may assume that $\omega_1/\omega_2 \in \mathfrak{H}$, by exchanging ω_1 and ω_2 if necessary. In this setting, we have:

Theorem 2. *The maximal unramified abelian extension of* K *can be generated by* $j(\omega_1/\omega_2)$ *over* K.

This is the first main theorem of the classical theory of complex multiplication. To construct ramified abelian extensions of K, one needs modular functions of higher level (see below) or elliptic functions with periods ω_1, ω_2. Even Th. 2 can fully be understood with the knowledge of elliptic functions or elliptic curves, though such are not explicitly involved in the statement. Therefore, our next task is to recall some elementary facts on this subject. But before that, it will be worth discussing a few elementary facts about the fractional linear transformations and discontinuous groups.

Every $\alpha = \begin{pmatrix} a & b \\ c & d \end{pmatrix} \in GL_2(C)$ acts on the Riemann sphere $C \cup \{\infty\}$ by the rule (2.1). With a suitable element ξ of $GL_2(C)$, $\beta = \xi\alpha\xi^{-1}$

has one of the following two normalized forms:

(i) $\beta(z) = z + \lambda$,

(ii) $\beta(z) = \kappa z$,

with constants λ and κ. This can be shown, for example, by taking the Jordan form of α. In the first case, we call α parabolic; in the second case, α is called elliptic, hyperbolic, or loxodromic, according as $|\kappa| = 1$, κ real, or otherwise. In this classification, we exclude the identity transformation, which is represented by the scalar matrices.

If $\alpha \in GL_2(R)$ and $\det(\alpha) > 0$, α maps $\mathfrak{H}$ onto itself, and α is

elliptic if α has exactly one fixed point in $\mathfrak{H}$,
hyperbolic if α has two fixed points in $R \cup \{\infty\}$,
parabolic if α has only one fixed point in $R \cup \{\infty\}$.

No transformation in $GL_2(R)$ with positive determinant is loxodromic.

If we put

$$SO_2(R) = \{\alpha \in SL_2(R) \mid \alpha \cdot {}^t\alpha = 1_2\},$$

then it can easily be verified that $SO_2(R)$ is the set of all elements of $SL_2(R)$ which leave the point i fixed. Therefore the map

$$SL_2(R) \ni \alpha \mapsto \alpha(i) \in \mathfrak{H}$$

gives a diffeomorphism of the quotient $SL_2(R)/SO_2(R)$ onto $\mathfrak{H}$. It is a fruitful idea to regard $\mathfrak{H}$ as such a quotient. But I shall not pursue this view point, from which one can actually start investigation in various directions.

We see easily that a differential form

$$y^{-2} dx \wedge dy \qquad (z = x + iy)$$

on $\mathfrak{H}$ is invariant under $SL_2(R)$. Therefore we can introduce an invariant measure on $\mathfrak{H}$ by means of this form.

To speak of an automorphic function for a Γ with non-compact $\mathfrak{H}/\Gamma$, we have to introduce the notion of cusp. Let Γ be a discrete subgroup of $SL_2(R)$. We call a point s of $R \cup \{\infty\}$ a <u>cusp of</u> Γ if there exists a parabolic element γ of Γ leaving s fixed. Let

$$\Gamma_s = \{\alpha \in \Gamma \mid \alpha(s) = s\}.$$

Then one can find an element ρ of $SL_2(R)$ so that $\rho(\infty) = s$, and $\rho\Gamma_s\rho^{-1}$ is generated by $\begin{pmatrix} 1 & 1 \\ 0 & 1 \end{pmatrix}$ and possibly by -1_2. Then we define an <u>automorphic function on</u> $\mathfrak{H}$ <u>with respect to</u> Γ to be a meromorphic function on $\mathfrak{H}$ satisfying the following two conditions:

(i) $f(\gamma(z)) = f(z)$ <u>for all</u> $\gamma \in \Gamma$.

(ii) <u>If</u> s <u>is a cusp of</u> Γ <u>and</u> ρ <u>is as above, then</u> $f(\rho(z))$ <u>is a meromorphic function in</u> $q = e^{2\pi i z}$ <u>in a neighborhood of</u> $q = 0$.

(Here note that if f satisfies (i), then $f(\rho(z))$ is invariant under $z \mapsto z+1$, hence $f(\rho(z))$ is always meromorphic at least in the domain $0 < |q| < r$ for some $r > 0$.)

Let $\mathfrak{H}^*$ be the join of $\mathfrak{H}$ and all the cusps of Γ. Then Γ acts on $\mathfrak{H}^*$. One can define a structure of Riemann surface on $\mathfrak{H}^*/\Gamma$ by taking $\exp[2\pi i . \rho^{-1}(z)]$ as a local parameter around the point s. (Actually the proof of the fact that $\mathfrak{H}^*/\Gamma$ is a Hausdorff space is not difficult, but non-trivial.) Then an automorphic function with respect to Γ, defined above, is nothing else than a meromorphic function on the Riemann surface $\mathfrak{H}^*/\Gamma$, regarded as a function on $\mathfrak{H}$. The above discussion about $SL_2(Z)$ is a special case of these facts. Now the following facts are known:

<u>Proposition 1.</u> <u>$\mathfrak{H}^*/\Gamma$ is compact if and only if $\mathfrak{H}/\Gamma$ has a finite measure with respect to the above invariant measure.</u>

<u>Proposition 2.</u> <u>Suppose that $\mathfrak{H}/\Gamma$ has a finite measure. Then $\mathfrak{H}/\Gamma$ is compact if and only if Γ has no parabolic element.</u>

As for elliptic elements, the following proposition holds:

<u>Proposition 3.</u> <u>Let z be a point of $\mathfrak{H}$ fixed by an elliptic element of Γ. Let $\Gamma_z = \{\alpha \in \Gamma \mid \alpha(z) = z\}$. Then Γ_z is a cyclic group of finite order.</u>

Such a point z is called an <u>elliptic point of</u> Γ, and the order of $\Gamma_z \cdot \{\pm 1_2\}/\{\pm 1_2\}$ is called the <u>order</u> of the point z (with respect to Γ). Two elliptic points or cusps are called <u>equivalent</u> if they are transformed to each other by elements of Γ. If $\mathfrak{H}/\Gamma$ is of finite measure, there are only a finite number of inequivalent elliptic points and cusps, and the following formula holds:

$$\frac{1}{2\pi}\iint_{\mathfrak{H}/\Gamma} y^{-2}dx\,dy = 2g - 2 + h + \Sigma_z (1 - 1/e_z). \tag{2.3}$$

Here g is the genus of $\mathfrak{H}^*/\Gamma$; h is the number of inequivalent cusps; Σ_z is the sum extended over all inequivalent elliptic points; e_z is the order of z. For $\Gamma = SL_2(Z)$, one has $g = 0$, $h = 1$, $e_z = 2$ or 3 according as $z = \sqrt{-1}$ or $z = (-1 + \sqrt{-3})/2$.

For every positive integer N, set

$$\Gamma(N) = \{\alpha \in SL_2(Z) \mid \alpha \equiv 1_2 \bmod N.M_2(Z)\}.$$

An automorphic function with respect to $\Gamma(N)$ is called a <u>modular function of level</u> N.

3. Elliptic curves and the fundamental theorems of the classical theory of complex multiplication

Let L be a lattice in the complex plane, i.e., a free Z-submodule of C of rank 2 which is discrete. Then C/L is a compact Riemann surface of genus one. An elliptic function with periods in L is a meromorphic function on C invariant under the translation $u \mapsto u + \omega$ for every $\omega \in L$. Define complex numbers g_2, g_3 and meromorphic functions $\wp(u)$ and $\wp'(u)$ on C by

(3.1)
$$g_2 = g_2(L) = 60\,\Sigma\omega^{-4},$$
$$g_3 = g_3(L) = 140\,\Sigma\omega^{-6},$$

$$\wp(u) = \wp(u;\,L) = u^{-2} + \Sigma[(u-\omega)^{-2} - \omega^{-2}],$$
$$\wp'(u) = \wp'(u;\,L) = -2\Sigma(u-\omega)^{-3},$$

where Σ denotes the sum extended over all non-zero ω in L. Then it is well-known that

(3.2)
$$\wp'(u)^2 = 4\wp(u)^3 - g_2\wp(u) - g_3;$$

(3.3) *The field of all elliptic functions with periods in* L *coincides with* $C(\wp, \wp')$, *the field generated by* $\wp$ *and* $\wp'$ *over* C.

Now let E be the algebraic curve defined by

(3.4)
$$y^2 = 4x^3 - g_2x - g_3.$$

Here we consider E as the set of all points (x, y) satisfying (3.4) with x, y in C, together with a point (∞, ∞). Then the map

$$C \ni u \mapsto (\wp(u), \wp'(u)) \in E$$

gives a holomorphic isomorphism of C/L onto E in the sense of complex manifold. It is also known that any elliptic curve (i.e. an algebraic curve of genus one) defined over C is isomorphic to a curve of this type, and hence to a complex torus.

Take a basis $\{\omega_1, \omega_2\}$ of L over Z. We may assume that $\omega_1/\omega_2 \in \mathfrak{H}$. Then one can easily show that

$$\omega_1/\omega_2 \longleftrightarrow E$$

defines a <u>one-to-one correspondence between</u> $\mathfrak{H}/\Gamma$ <u>and all the isomorphism-classes of elliptic curves</u>. Furthermore we have an important relation

$$j(\omega_1/\omega_2) = g_2^3/(g_2^3 - 27g_3^2). \tag{3.5}$$

One should note that the right hand side can be obtained purely algebraically from the defining equation (3.4) for E, while the left is defined analytically. This coincidence of an algebraic object with an analytic object has a deep meaning, though we know, from (3.1), that g_2 and g_3 depend analytically on ω_1 and ω_2. We call the number expressed by (3.5) <u>the invariant of</u> E. Two elliptic curves have the same invariant if and only if they are isomorphic.

Let us now observe that any holomorphic endomorphism of $E = C/L$ is obtained by $u \mapsto \lambda u$ with a complex number λ satisfying

$\lambda . L \subset L$. Let End(E) denote the ring of all such endomorphisms. It can easily be proved that End(E) is isomorphic to Z unless $Q(\omega_1/\omega_2)$ is an imaginary quadratic field. Assume that $Q(\omega_1/\omega_2)$ is imaginary quadratic, and put $K = Q(\omega_1/\omega_2)$. Then End(E) is isomorphic to a subring of the ring $\mathfrak{O}$ of all algebraic integers in K, which generates K. In this case we say that E <u>has complex multiplications</u>. In particular, if $L = Z\omega_1 + Z\omega_2$ is an ideal in K, End(E) is isomorphic to $\mathfrak{O}$. Put $j_o = j(\omega_1/\omega_2)$. For a given L (or ω_1, ω_2), one can find the equation (3.4) so that g_2 and g_3 are contained in $Q(j_o)$. Moreover j_o is an algebraic number if E has complex multiplications.

Now write E as $E(\mathfrak{a})$ if $L = \mathfrak{a}$ for an ideal $\mathfrak{a}$ in K. We choose the equation for $E(\mathfrak{a})$ so that $g_2, g_3 \in Q(j_o)$. Suppose we could somehow prove that $K(j_o)$ is an abelian extension of K. (Anyway this is not the most difficult point of the theory.) Take a prime ideal $\mathfrak{p}$ in K unramified in $K(j_o)$, and let $\sigma = [\mathfrak{p}, K(j_o)/K]$ (= the Frobenius automorphism of $K(j_o)$ over K for $\mathfrak{p}$). Then g_2 and g_3 are meaningful. Therefore we can define an elliptic curve $E(\mathfrak{a})^\sigma$ by

$$E(\mathfrak{a})^\sigma : y^2 = 4x^3 - g_2^\sigma x - g_3^\sigma .$$

Then one has a fundamental relation:

(3.6) $E(\mathfrak{a})^\sigma$ <u>is isomorphic to</u> $E(\mathfrak{a}\mathfrak{p}^{-1})$.

If we denote by $j(\mathfrak{a})$ the invariant of $E(\mathfrak{a})$, then (3.6) is equivalent to

(3.7) $$j(\mathfrak{a})^{\sigma} = j(\mathfrak{a}\mathfrak{p}^{-1}).$$

From the relation (3.6) or (3.7), one can easily derive Th. 2 and also the reciprocity law in the extension $K(j_0)$ of K. Here I do not go into detail of the proof of (3.6), but would like to call the reader's attention to the following point: Although no elliptic curves appear in Th. 2, they conceal themselves in it through the above (3.6) and the following facts.

(3.8) <u>The quotient $\mathfrak{H}/\Gamma$ is in one-to-one correspondence with all the isomorphism classes of elliptic curves.</u>

(3.9) $j(\omega_1/\omega_2)$ <u>is the invariant of an elliptic curve E isomorphic to</u> $C/(Z\omega_1 + Z\omega_2)$.

(3.10) <u>If</u> $Q(\omega_1/\omega_2)$ <u>is imaginary quadratic,</u> $\mathrm{End}(E)$ <u>is non-trivial.</u>

Let us now consider the question of generalizing Theorems 1 and 2 to the fields of higher degree. We observe that there are three objects:

(A) elliptic curve,

(B) modular function,

(C) imaginary quadratic field.

Among many possible ideas, one can take the most naive one, namely ask whether there exist generalizations of (A), (B), (C) whose interrelation is similar to that of the original ones, as described in (3.8-10).

The answer is affirmative but not unique. It may be said that the world of mathematics is built with a great harmony but not always in the form which we expect before unveiling it. This certainly applies to our present question. I shall, however, first present a comparatively simple answer which consists of the following three objects:

(A') abelian variety,

(B') Siegel modular function,

(C') totally imaginary quadratic extension of a totally real algebraic number field.

At least this will include the above result concerning elliptic curves as a special case. A different type of theory, which I feel rather unexpected, and which also generalizes Th. 2, will be discussed later.

4. Relation between the points of finite order on an elliptic curve and the modular functions of higher level.

Before talking about abelian varieties, let us discuss the topic given as the title of this section. Any hasty reader may skip this section, and come back afterward.

Fix a positive integer N. Observe that any point t on E such that $Nt = 0$ can be expressed as

$$t = (\wp((a\omega_1 + b\omega_2)/N),\ \wp'((a\omega_1 + b\omega_2)/N))$$

with integers a, b. Now, for each ordered pair (a, b) of integers such that $(a, b) \not\equiv (0, 0) \bmod (N)$, we can define a meromorphic function $f^N_{ab}(z)$ on $\mathfrak{H}$ by

$$f^N_{ab}(z) = \frac{g_2(L)}{g_3(L)} \cdot \wp\left(\frac{a\omega_1 + b\omega_2}{N}; L\right),$$

where $z = \omega_1/\omega_2$ and $L = Z\omega_1 + Z\omega_2$. This is possible because the right hand side depends only on $z = \omega_1/\omega_2$. Then

$$f^N_{ab}(z) = f^N_{cd}(z) \Longleftrightarrow \quad (a, b) \equiv (c, d) \bmod (N) \quad \text{or} \quad (a, b) \equiv (-c, -d) \bmod (N).$$

By a simple calculation, we can show that, for every $\alpha \in SL_2(Z)$,

$$f^N_{ab}(\alpha(z)) = f^N_{cd}(z) \text{ if } (a \ \ b)\alpha = (c \ \ d)$$

Therefore, $f^N_{ab}(z) = f^N_{ab}(\alpha(z))$ for all (a, b) if and only if α belongs to $\Gamma(N).\{\pm 1_2\}$. It follows that j and the f^N_{ab}, for all (a, b), generate the field of all modular functions of level N. Roughly speaking, the modular functions of level N can be obtained from the invariant of elliptic curves and points of order N on the curves. Now we have the following result which is an analogue of Th. 1 for an imaginary quadratic field.

Theorem 3. *Let K be as above, and $\mathfrak{a}$ an ideal in K. Take $\mathfrak{a}$ as L, and let $\mathfrak{a} = Z\omega_1 + Z\omega_2$ with $\omega_1/\omega_2 \in \mathfrak{H}$. Suppose that $g_2(\mathfrak{a})g_3(\mathfrak{a}) \neq 0$. Then the maximal abelian extension of K is generated over K by $j(\omega_1/\omega_2)$ and the $f^N_{ab}(\omega_1/\omega_2)$ for all N, a, b, with a fixed $\mathfrak{a}$.*

It should be observed here that $f^N_{ab}(\omega_1/\omega_2)$ is a special value of an elliptic function and a special value of a modular function of level N as well. This coincidence will not necessarily be retained in one of our later generalizations.

We note that $g_2(\mathfrak{a}) = 0$ or $g_3(\mathfrak{a}) = 0$ according as $K = Q((-1 + \sqrt{-3})/2)$ or $K = Q(\sqrt{-1})$. In these special cases, we can still obtain the same type of result by modifying the definition of f^N_{ab} suitably.

The function field $C(j, f^N_{ab})$, with a fixed N, is a Galois extension of $C(j)$ whose Galois group is isomorphic to $\Gamma(1)/\Gamma(N)\cdot\{\pm 1_2\}$; the latter group is isomorphic to $SL_2(Z/NZ)/\{\pm 1_2\}$. Since our purpose is to construct number fields by special values of functions, it is meaningful to take Q, instead of C, as the basic field. Then we obtain:

Theorem 4. *$Q(j, f^N_{ab})$ is a Galois extension of $Q(j)$ whose Galois group is isomorphic to $GL_2(Z/NZ)/\{\pm 1_2\}$, and the following statements hold.*

(i) *For every $\alpha \in GL_2(Z/NZ)$, the action of α as an element of the Galois group is given by $f^N_{ab} \mapsto f^N_{cd}$ with $(c\ d) = (a\ b)\alpha$.*

(ii) *If $\gamma \in SL_2(Z)$, the action of γ mod (N) on $Q(j, f^N_{ab})$ is obtained by $\varphi(z) \mapsto \varphi(\gamma(z))$ for $\varphi \in Q(j, f^N_{ab})$.*

(iii) $Q(j, f_{ab}^{N})$ contains $\zeta = e^{2\pi i/N}$, and $\alpha \in GL_2(Z/NZ)$ sends ζ to $\zeta^{\det(\alpha)}$.

We shall later extend this theorem to the field of automorphic functions with respect to a more general type of group.

5. Abelian varieties and Siegel modular functions

A non-singular projective variety of dimension n, defined over C, is called an abelian variety if it is, as a complex manifold, isomorphic to a complex torus of dimension n. An elliptic curve is nothing but an abelian variety of dimension one. We know that any one dimensional complex torus defines an elliptic curve, but such is not true in the higher dimensional case. To explain the necessary condition, let L be a lattice in the n-dimensional complex vector space C^n, i.e., a discrete free Z-submodule of rank $2n$ in C^n. Then the complex torus C^n/L has a structure of projective variety, and hence becomes an abelian variety, if and only if there exists an R-valued R-bilinear form $\mathfrak{E}(x, y)$ on C^n with the following properties:

(R_1) $\mathfrak{E}(x, y) = -\mathfrak{E}(y, x)$;

(R_2) $\mathfrak{E}(x, y) \in Z$ if $x, y \in L$;

(R_3) $\mathfrak{E}(x, \sqrt{-1}y)$ is symmetric in (x, y), and $\mathfrak{E}(x, \sqrt{-1}x) > 0$ for all $x \neq 0$.

We call such $\mathcal{E}$ a Riemann form on C^n/L. Take a basis $\{v_1, \ldots, v_{2n}\}$ of L over Z, and regard the elements of C^n as olumn vectors. Then we obtain a matrix

$$\Omega = (v_1 \quad v_2 \quad \ldots \quad v_{2n})$$

f $n \times 2n$ type, which may be called a period matrix for C^n/L. Dene a matrix $P = (p_{ij})$ of size 2n by $p_{ij} = \mathcal{E}(v_i, v_j)$. Then the bove (R_{1-3}) are equivalent to the following (R'_{1-3}):

(R'_1) ${}^tP = -P$;

(R'_2) $p_{ij} \in Z$;

(R'_3) $\Omega P^{-1} \cdot {}^t\Omega = 0$, and $-\sqrt{-1}\Omega P^{-1} \cdot {}^t\bar{\Omega}$ is a positive definite hermitian matrix.

he matrix P (or its inverse) is called a principal matrix of Ω.

Assuming these conditions, let A be a projective variety isoorphic to C^n/L. Shifting the law of addition of C^n/L to A, we an define a structure of commutative group on A. Then the map

$$A \times A \ni (x, y) \mapsto x + y \in A$$

an be expressed rationally by the coordinates of x and y. This is lassically known as the addition theorem of abelian functions.

In general, a projective variety A, defined over any field of ny characteristic, is called an abelian variety, if there exist rational appings $f: A \times A \to A$ and $g: A \to A$ which define a group strucre on A by $f(x, y) = x + y$, $g(x) = -x$. Additive notation is used since any such group structure on a projective variety can be shown to be commutative. It should be observed that such a variety defined over C, being a connected compact commutative complex Lie group, must be isomorphic to a complex torus.

If $n = 1$, there is a single universal family of elliptic curves parametrized by the point of $\mathfrak{h}$. If $n > 1$, however, there are infinitely many families of abelian varieties depending on the elementary divisors of P, as shown in the Supplement below. But we shall first fix our attention to one particular family by considering only abelian varieties for which $P = J_n$, where

$$J_n = \begin{bmatrix} 0 & -1_n \\ 1_n & 0 \end{bmatrix} . \tag{5.1}$$

Under this assumption, let ω_1 and ω_2 be the square matrices of size n composed of the first and the last n columns of Ω respectively. One can show that ω_2 is invertible. Put $z = \omega_2^{-1}\omega_1$. If we change the coordinate system of C^n by ω_2, we may assume that Ω is of the form

$$\Omega = (z \;\; 1_n) .$$

Now it can be shown (see Supplement below) that z is symmetric and $\mathrm{Im}(z)$ is positive definite. We denote by $\mathfrak{h}_n$ the set of all such z of degree n. Thus every abelian variety, under the assumption that P has the form (5.1), corresponds to a point of $\mathfrak{h}_n$, though z is not unique for a given abelian variety. Moreover, every point of $\mathfrak{h}_n$

corresponds to such an abelian variety. Obviously $\mathfrak{H}_1 = \mathfrak{H}$.

As an analogue of $SL_2(R)$, we introduce a group

$$Sp(n, R) = \{U \in SL_{2n}) | {}^tUJ_nU = J_n\}.$$

For every $U = \begin{bmatrix} a & b \\ c & d \end{bmatrix} \in Sp(n, R)$ with a, b, c, d in $M_n(R)$, we define the action of U on $\mathfrak{H}_n$ by

$$U(z) = (az + b)(cz + d)^{-1}.$$

Put

$$Sp(n, Z) = SL_{2n}(Z) \cap Sp(n, R).$$

When $n > 1$, we can define an <u>automorphic function with respect to</u> $Sp(n, Z)$ to be a meromorphic function on $\mathfrak{H}_n$ invariant under $Sp(n, Z)$. Fortunately, if $n > 1$, it is not necessary to impose any further condition like that we needed in the case $n = 1$. Such a function is usually called a <u>Siegel modular function</u> (of degree n and level one).

Put $\Gamma = Sp(n, Z)$. Now one can ask whether the quotient $\mathfrak{H}_n/\Gamma$ is in one-to-one correspondence with all the isomorphism classes of abelian varieties of type (5.1). This is almost so but not quite. To get an exact answer, we define 2n real coordinate functions $x_1(u)$, ..., $x_{2n}(u)$ $(u \in C^n)$ by $u = \Sigma_{i=1}^n x_i(u)v_i$, and consider a cohomology class c on A represented by a differential form

$$q = \Sigma_{i<j}\, p_{ij}\, dx_i \wedge dx_j$$

of degree 2. Such a c is called a *polarization* of A, and the structure (A, c) formed by A and its polarization c is called a *polarized abelian variety. Then* $\mathfrak{H}_n/\Gamma$ *represents all the isomorphism classes of polarized abelian varieties of type* (5.1), the isomorphism being defined in a natural way.

Our next question is about the existence of some functions similar to j and the analogue of (3.5). First one should note that *there exists a Zariski open subset* V *of a projective variety* V* *and a holomorphic mapping* φ *of* $\mathfrak{H}_n$ *into* V *which induces a biregular isomorphism of* $\mathfrak{H}_n/\Gamma$ *onto* V. This was proved by W. L. Baily using the Satake compatification of $\mathfrak{H}_n/\Gamma$. We call such a couple (V, φ) a *model* for $\mathfrak{H}_n/\Gamma$. This is not sufficient for our purpose. In fact, in the case n = 1, the function $(\alpha j + \beta)/(\gamma j + \delta)$ plays a role of φ, for any $\begin{bmatrix} \alpha & \beta \\ \gamma & \delta \end{bmatrix} \in GL_2(C)$. Of course one can not replace j by such a function in Th. 2. Furthermore, we would like to have an analogue of (3.5). Therefore a further refinement is necessary, and can be given as follows:

Theorem 5. *There exists a couple* (V, φ) *with the following properties:*

(i) (V, φ) *is a model for* $\mathfrak{H}_n/\Gamma$.

(ii) V *is defined over* Q.

(iii) *Let* (A, c) *be a polarized abelian variety with a* P *of type* (5.1), *defined over a subfield* k *of* C, *and* σ *an isomorphism of* k *into* C. *Let* z *and* z' *be points on* $\mathfrak{H}_n$ *corresponding to* (A, c)

<u>and</u> $(A, c)^\sigma$ <u>respectively</u>. <u>Then the coordinates of the point</u> $\varphi(z)$ <u>belong to</u> k, <u>and</u> $\varphi(z)^\sigma = \varphi(z')$.

In (iii), we of course consider A as a projective variety defined by some homogeneous equations. Now one can prove that the cohomology class c is represented by a divisor on A (i.e. an $(n-1)$-dimensional algebraic subset of A). If the defining equations for A and such a divisor have coefficients in a field k, we say that (A, c) is defined over k. If σ is as in (iii), the transforms of the equations by σ define an abelian variety together with a divisor, which turns out to be a polarized abelian variety of type (5.1), which we write as $(A, c)^\sigma$. We can actually prove a stronger statement than (iii), which is roughly as follows:

(iv) <u>If</u> (A', c') <u>is a specialization of</u> (A, c) <u>over</u> Q, <u>and</u> z' <u>corresponds to</u> (A', c'), <u>then</u> $((A', c'), \varphi(z'))$ <u>is a specialization of</u> $((A, c), \varphi(z))$ <u>over</u> Q.

For details we refer the reader to the paper [20] in §12.

Thus φ plays a role similar to j. It is analytic on $\mathfrak{H}_n$, and at the same time, it is a rational expression of the coefficients of defining equations for (A, c), as explained in (iii). From (i) it follows that the coordinates of $\varphi(z)$ generate the whole field of Siegel modular functions of degree n. The couple (V, φ) can be characterized by these properties (i, ii, iii). Namely, if (V', φ') is another couple with the same properties, there exists a biregular isomorphism f of V onto V' defined over Q such that $\varphi' = f \circ \varphi$. Moreover, from (iii), we see that the field $Q(\varphi(z))$ has an invariant meaning for the isomorphism class of (A, c). We call it _the field of moduli of_ (A, c).

Actually we can prove all these things without assuming $P = J_n$. For each choice of P (or rather for a choice of elementary divisors of P), one obtains a group Γ_P (see _Supplement_ below) acting on $\mathfrak{H}_n$ and a couple (V_P, φ_P) with the properties (i, ii, iii) modified suitably. Further, by considering the points of finite order on the abelian varieties, one can generate the field of automorphic functions with respect to congruence subgroups of $Sp(n, Z)$; one then obtains a theorem analogous to Th. 4.

The next thing to do is the investigation of special members of our family of abelian varieties, analogous to elliptic curves with complex multiplications. This will be done in §6.

Supplement to §5. To discuss the families of abelian varieties of a more general type, for which P is not necessarily of the form (5.1), first we recall a well known

Lemma. Let P be an invertible alternating matrix of size 2n _with entries in_ Z. _Then there exists an element_ U _of_ $GL_{2n}(Z)$ _such that_

$$ {}^tUPU = \begin{bmatrix} 0 & -e \\ e & 0 \end{bmatrix}, \qquad e = \begin{bmatrix} e_1 & & \\ & \ddots & \\ & & e_n \end{bmatrix}, $$

where the e_i _are positive integers satisfying_ $e_{i+1} \equiv 0 \bmod (e_i)$.

Therefore, to discuss Ω satisfying (R'_{1-3}), we may assume that $P = \begin{bmatrix} 0 & -e \\ e & 0 \end{bmatrix}$ with e as in the above lemma. Let Y_P be the space of all such Ω, and let

$$B = \begin{bmatrix} 1_n & 0 \\ 0 & e \end{bmatrix}, \quad G_P = \left\{ U \in SL_{2n}(R) \mid {}^tUPU = P \right\}.$$

In particular $G_P = Sp(n, R)$ if $e = 1_n$. If $\Omega \in Y_P$ and $U \in G_P$, then $\Omega U \in Y_P$. Obviously ${}^tBJ_nB = P$, hence $BG_PB^{-1} = Sp(n, R)$. Now write $\Omega = (v \ v')$ with two elements v and v' of $M_n(C)$. Then, from (R'_3), we see easily that

$$ve^{-1} \cdot {}^tv' - v'e^{-1} \cdot {}^tv = 0,$$

$$\sqrt{-1}(v'e^{-1} \cdot {}^t\bar{v} - ve^{-1} \cdot {}^t\bar{v}') \text{ is positive definite.}$$

The last fact implies that v and v' are invertible. From these relations it follows that $ev'^{-1}v$ is symmetric and has a positive definite imaginary part, i.e., $ev'^{-1}v \in \mathfrak{H}_n$.

If $z \in \mathfrak{H}_n$ and $U = \begin{bmatrix} a & b \\ c & d \end{bmatrix} \in Sp(n, R)$, then $(z \ 1_n)U \in Y_{J_n}$, hence by the above result, $(z \ 1_n)U = \lambda(w \ 1_n)$ with $\lambda \in M_n(C)$ and $w \in \mathfrak{H}_n$. Then one obtains $w = (az + b)(cz + d)^{-1}$. This shows that the action of U on $\mathfrak{H}_n$ can actually be defined. Since the action of U^{-1} can be defined, U gives a holomorphic automorphism of $\mathfrak{H}_n$. Now set

$$\Gamma'_P = G_P \cap SL_{2n}(Z), \quad \Gamma_P = B\Gamma'_P B^{-1}.$$

It can easily be seen that Γ_P is a discrete subgroup of $Sp(n,R)$. Then $\mathfrak{H}_n/\Gamma_P$ <u>represents all the isomorphism classes of polarized abelian varieties with principal matrix</u> P.

The notion of polarization can also be defined in the case of positive characteristic. Given an abelian variety A defined over a field of any characteristic, and given a divisor X on A, let L be the linear space of all rational functions on A whose poles are contained in X (even with multiplicities). Take a basis $\{f_o, f_1, \ldots, f_N\}$ of L over k, and consider the map

$$A \ni x \mapsto (f_o(x), \ldots, f_N(x)) \in \text{projective } N\text{-space}.$$

We call X <u>ample</u> if this is a biregular embedding of A into the projective space. Now a <u>polarization</u> of A is a set $\mathfrak{X}$ of divisors on A satisfying the following conditions:

(1) $\mathfrak{X}$ <u>contains an ample divisor</u>.

(2) <u>If</u> X <u>and</u> X' <u>belong to</u> $\mathfrak{X}$, <u>then there are two positive integers</u> m <u>and</u> m' <u>such that</u> mX <u>and</u> $m'X'$ <u>are algebraically equivalent</u>.

(3) $\mathfrak{X}$ <u>is a maximal set satisfying the above two conditions</u>.

In general, two divisors X and Y are called <u>algebraically equivalent</u>, if there exist a divisor W and its specializations W_1 and W_2 over an algebraically closed field such that $X - Y = W_1 - W_2$. If the universal domain is C, then the algebraic equivalence of divisors coincides with the homological equivalence. Moreover, if a divisor X represents the cohomology class c obtained from a Riemann form, then $3X$ is ample. Every abelian variety, defined over a field of any characteristic, has a polarization, since it always has an ample divisor.

Now a <u>polarized abelian variety</u> is a couple $(A, \mathfrak{X})$ formed by an abelian variety A and a polarization $\mathfrak{X}$ of A. This definition is equivalent to the previous one, if the universal domain is C. An isomorphism of A of A' is called an <u>isomorphism</u> of $(A, \mathfrak{X})$ to $(A', \mathfrak{X}')$ if it sends $\mathfrak{X}$ to $\mathfrak{X}'$. For a given $(A, \mathfrak{X})$, we can prove that there exists a field k_o with the following properties:

(i) <u>If</u> $(A, \mathfrak{X})$ <u>is defined over</u> k, <u>then</u> k_o <u>is contained in</u> k.

(ii) <u>For an isomorphism</u> σ <u>of</u> k <u>into the universal domain,</u> $(A^\sigma, \mathfrak{X}^\sigma)$ <u>is isomorphic to</u> $(A, \mathfrak{X})$ <u>if and only if</u> σ <u>is the identity mapping on</u> k_o.

If the universal domain is C, k_o is uniquely determined by the isomorphism class of $(A, \mathfrak{X})$, and is called <u>the field of moduli of</u> $(A, \mathfrak{X})$. This of course coincides with $Q(\varphi(z))$ in the special case $P = J_n$.

6. The endomorphism-ring of an abelian variety; the field of moduli of an abelian variety with many complex multiplications

For an abelian variety A, we denote by $\mathrm{End}(A)$ the ring of all holomorphic endomorphisms of A. If A is isomorphic to a complex torus C^n/L, every endomorphism of A corresponds to an element T of $M_n(C)$, regarded as a C-linear transformation on C^n, satisfying $T(L) \subset L$. Therefore $\mathrm{End}(A)$ is a free Z-module of finite rank. Let $\mathrm{End}_Q(A) = \mathrm{End}(A) \otimes_Z Q$, and $W = Q.L$. Then W is a vector space over Q of dimension $2n$, which spans C^n over R, and $\mathrm{End}_Q(A)$ is isomorphic to the ring

$$\{ T \in M_n(C) \mid T(W) \subset W \} .$$

For each element of $\mathrm{End}_Q(A)$, consider the corresponding element T of $M_n(C)$. Then we get a faithful representation of $\mathrm{End}_Q(A)$ by complex matrices of size n, which we call *the complex representation of* $\mathrm{End}_Q(A)$. On the other hand, with respect to a basis of W over Q (for example, $\{v_1, \dots, v_{2n}\}$ considered in §5), we obtain a representation of $\mathrm{End}_Q(A)$ by rational matrices of degree $2n$, which we call *the rational representation of* $\mathrm{End}_Q(A)$. As an easy exercise of linear algebra, one can prove:

Lemma. The rational representation of $\mathrm{End}_Q(A)$ *is equivalent to the direct sum of the complex representation of* $\mathrm{End}_Q(A)$ *and its complex conjugate.*

Let k be a field of definition for A and the elements of $End(A)$, and let D be the vector space of all linear invariant differential forms on A, defined over k. If $z_1, \ldots, z_n$ are the complex coordinate functions in C^n, then $dz_1, \ldots, dz_n$ are considered as invariant differential forms on A, and one has

(6.1) $$D \otimes_k C = \Sigma_{i=1}^n C.\,dz_i \,.$$

Now every $\lambda \in End(A)$ acts on D as usual; denote the action by λ^*. Then $\lambda \mapsto \lambda^*$ can be extended to an anti-isomorphism of $End_Q(A)$ into the ring of linear transformations in D. From the relation (6.1), we obtain

(6.2) <u>This anti-isomorphism is equivalent to the transpose of the complex representation of</u> $End_Q(A)$.

Let $\mathfrak{E}$ be a Riemann form on C^n/L. For every $T \in End_Q(A)$, one can define an element T^ρ of $End_Q(A)$ by

(6.3) $$\mathfrak{E}(x, Ty) = \mathfrak{E}(T^\rho x, y).$$

Here we identify an element of $End_Q(A)$ with the corresponding element of $M_n(C)$. Then ρ <u>is a positive involutinn of</u> $End_Q(A)$. In general, an <u>involution</u> of an associative algebra S over Q (or R) is, by definition, a one-to-one map ρ of S onto S such that

$$(x+y)^\rho = x^\rho + y^\rho ,$$

$$(xy)^\rho = y^\rho x^\rho ,$$

$$(x^\rho)^\rho = x \qquad (x, y \in S).$$

Such a ρ is called <u>positive</u> if $\mathrm{Tr}(xx^\rho) > 0$ for $0 \neq x \in S$, where Tr denotes the trace of a regular representation of S over Q.

If an algebra S over Q or R has a positive involution ρ, then S has no nilpotent ideal other than $\{0\}$. In fact, if $x, \neq 0$, belongs to a nilpotent ideal, then $\mathrm{Tr}(xy) = 0$ for every $y \in S$, but this is a contradiction, since $\mathrm{Tr}(xx^\rho) > 0$. It follows that S is semi-simple. If e is the identity element of a simple component of S, then $ee^\rho \neq 0$, hence $e^\rho = e$. It follows that ρ is stable on each simple component of S. Thus the classification of S and ρ can be reduced to the case of simple algebras.

If S is an algebra over Q with a positive involution ρ, we can extend ρ to a positive involution of $S \otimes_Q R$. In particular, consider the case where S is an algebraic number field, and use the letter K instead of S. Put

$$F = \{x \in K \mid x^\rho = x\}.$$

Then $[K : F] = 1$ or 2. By the general principle we just mentioned, ρ is extended to a positive involution of the tensor product $K \otimes_Q R$ which is a direct sum of copies of R or C. From this fact it is easy to see that the direct factors of $F \otimes_Q R$ are all real, i. e., F is totally real. Further, if $[K : F] = 2$, the direct factors of $K \otimes_Q R$ are all C, i. e., K is totally imaginary.

Conversely, let F be a totally real algebraic number field, K a totally imaginary quadratic extension of F, and ρ the non-trivial automorphism of K over F. Then ρ is a positive involution of K.

We fix such F, K, ρ, and consider a triple (A, c, θ) formed by a polarized abelian variety (A, c) and an isomorphism θ of K into $\mathrm{End}_Q(A)$ such that the map $\theta(a) \mapsto \theta(a^\rho)$ is exactly the restriction of the involution of $\mathrm{End}_Q(A)$ obtained as above. (Note that $\mathrm{End}_Q(A)$ may be larger than $\theta(K)$.) We assume also that $\theta(1)$ is the identity of $\mathrm{End}_Q(A)$. Take C^n/L and W as above. Then W may be regarded as a vector space over K, by means of the action of $\theta(K)$. Let m be the dimension of W over K, and $g = [F : Q]$. Then we have obviously

$$n = gm. \tag{6.4}$$

Now restrict the complex representation of $\mathrm{End}_Q(A)$ to $\theta(K)$. Then we obtain a representation Φ of K by complex matrices of size n. In this situation, we say that (A, c, θ) <u>is of type</u> (K, Φ). Since K is a field, Φ is equivalent to the direct sum of n isomorphisms of K into C. By our choice of K, there are exactly $2g$ isomorphisms of K into C, which can be written as

$$\tau_1, \ldots, \tau_g, \rho\tau_1, \ldots, \rho\tau_g \tag{6.5}$$

with a suitable choice of g isomorphisms $\tau_1, \ldots, \tau_g$ among them. Let r_ν and s_ν be the multiplicity of τ_ν and $\rho\tau_\nu$ in Φ, respectively, or symbolically, put

$$(6.6) \qquad \Phi \sim \Sigma_{\nu=1}^{g}(r_\nu \cdot \tau_\nu + s_\nu \cdot \rho\tau_\nu).$$

Note that $a^{\rho\sigma}$ is the complex conjugate of a^σ for every $a \in K$ and every isomorphism σ of K into C. From the above lemma it follows that $\Sigma_{\nu=1}^{g}(r_\nu + s_\nu).(\tau_\nu + \rho\tau_\nu)$ is equivalent to a rational representation of K. Therefore we have

$$r_1 + s_1 = r_2 + s_2 = \ldots = r_g + s_g .$$

Since Φ is of degree n and $n = mg$, we have

$$(6.7) \qquad r_\nu + s_\nu = m \qquad (\nu = 1, \ldots, g) .$$

In particular, if $m = 1$ (and hence $n = g$), either r_ν or s_ν is 0. Exchanging τ_ν and $\rho\tau_\nu$ if necessary, we may assume that

$$r_1 = \ldots = r_g = 1, \; s_1 = \ldots = s_g = 0,$$

i.e., $\Phi \sim \Sigma_{\nu=1}^{g}\tau_\nu$.

For a given K and $\tau_1, \ldots, \tau_g$, the existence of (A, c, θ) of type (K, Φ), with $\Phi \sim \Sigma_{\nu=1}^{g}\tau_\nu$, can be shown as follows. For every $s \in K$, let $u(s)$ denote the element of C^g with components

$s^{\tau_1}, \ldots, s^{\tau_g}$. Take any free Z-submodule $\mathfrak{a}$ of K of rank 2g. Put

$$(6.8) \qquad L = \{u(s) \mid s \in \mathfrak{a}\}.$$

It can easily be shown that L is a lattice in C^g, so that C^g/L is a complex torus. Take an element ζ of K so that

$$\zeta^\rho = -\zeta, \ \mathrm{Im}(\zeta^{\tau_\nu}) > 0 \ \text{for} \ \nu = 1, \ldots, g.$$

Define an R-valued alternating form $\mathcal{E}(x, y)$ on C^g by

$$(6.9) \qquad \mathcal{E}(x, y) = t.\Sigma_{\nu=1}^{g} \zeta^{\tau_\nu}(x_\nu \bar{y}_\nu - \bar{x}_\nu y_\nu)$$

where x_ν(resp. y_ν) is the νth component of x (resp. y), and t is a positive integer. For a suitable choice of t, we see easily that $\mathcal{E}$ becomes a Riemann form on C^g/L, hence C^g/L is isomorphic to an abelian variety A. From $\mathcal{E}$ we obtain a polarization c of A. For every $a \in K$, the diagonal matrix with diagonal elements $a^{\tau_1}, \ldots, a^{\tau_g}$ defines an element of $\mathrm{End}_Q(A)$, which we write $\theta(a)$. In particular, if $a\mathfrak{a} \subset \mathfrak{a}$, the matrix sends L into L, hence $\theta(a) \in \mathrm{End}(A)$. Thus we obtain (A, c, θ) of type (K, Φ). Actually one can prove that any (A, c, θ) of type (K, Φ) is constructed in this way. If $\mathfrak{a}$ is a fractional ideal in K, and $\mathfrak{o}$ denotes the ring of algebraic integers in K, then $\theta(\mathfrak{o}) \subset \mathrm{End}(A)$. If n = 1, our

(A, c, θ) is nothing but an elliptic curve isomorphic to $C/\mathfrak{a}$ (provided that τ_1 is the identity map of K).

Now taking a period matrix for A, we obtain a point z of $\mathfrak{H}_n$ as in §5. Here we assume that (A, c) is such that $P = J_n$, and $\theta(\mathfrak{O}) \subset \mathrm{End}(A)$. Let (V, φ) be a couple as in Theorem 5. Let $K(\varphi(z))$ be the field generated over K by the coordinates of the point $\varphi(z)$. One may naturally ask a question:

Is $K(\varphi(z))$ the maximal unramified abelian extension of K?

This is so if $n = 1$, as asserted by Theorem 2. But if $n > 1$, this is not necessarily true. To construct the maximal unramified abelian extension of K, we shall later discuss a function which is rather different from φ. However, even though φ is not a function with the expected property, $\varphi(z)$ has still an interesting number theoretical property, which is roughly described as follows:

Theorem 6. Let K' be the field generated over Q by $\Sigma_{\nu=1}^{g} a^{\tau_\nu}$ for all $a \in K$. Then $K'(\varphi(z))$ is an unramified abelian extension of K'.

It can be shown that K' is also a totally imaginary quadratic extension of a totally real algebraic number field. Obviously $K' = K$ if $n = 1$. However, both cases $K = K'$ and $K \neq K'$ can happen if $n > 1$. Even the degrees of K and K' over Q may be different. The field $K'(\varphi(z))$ is not necessarily the *maximal* unramified abelian extension of K'. Then how big is $K'(\varphi(z))$? We shall answer this question in the following section.

7. The class-field-theoretical characterization of $K'(\varphi(z))$

Let us first recall the fundamental theorems of class field theory. On this topic, I shall give an exposition which is somewhat out of mode, since such will be most convenient to describe the field $K'(\varphi(z))$.

Let F be an algebraic number field of finite degree, $\mathfrak{t}$ an integral ideal in F, and $\mathfrak{u}$ a (formal) product of real archimedean primes of F. For an element a of F, we write

$$a \equiv 1 \bmod^* \mathfrak{t}\mathfrak{u}$$

if there exist two algebraic integers b and c in F such that $a = b/c$, $b \equiv c \equiv 1 \bmod \mathfrak{t}$, and b, c are positive for every archimedean prime involved in $\mathfrak{u}$. We denote by $I(F, \mathfrak{t})$ the group of all fractional ideals in F prime to $\mathfrak{t}$, and by $P(F, \mathfrak{t}\mathfrak{u})$ the subgroup of $I(F, \mathfrak{t})$ consisting of all principal ideals (a) such that $a \equiv 1 \bmod^* \mathfrak{t}\mathfrak{u}$.

Let M be a finite abelian extension of F. For every prime ideal $\mathfrak{p}$ in F unramified in M, the Frobenius automorphism $[\mathfrak{p}, M/F]$ is meaningful. Let $\mathfrak{d}$ be the relative discriminant of M over F. Then we can define $[\mathfrak{a}, M/F]$ for every $\mathfrak{a} \in I(F, \mathfrak{d})$ so that

$$\mathfrak{a} \mapsto [\mathfrak{a}, M/F] \tag{7.1}$$

is a homomorphism of $I(F, \mathfrak{d})$ into the Galois group $G(M/F)$ of M over F. We have now

Theorem 7. *The map* (7.1) *is surjective, and its kernel contains* $P(F, \mathfrak{d}\mathfrak{u})$ *for some* $\mathfrak{u}$.

Therefore, if Y is the kernel, $G(M/F)$ is isomorphic to $I(F, \mathfrak{d})/Y$. Moreover, a prime ideal $\mathfrak{p}$ in F is fully decomposed in M if and only if $\mathfrak{p} \in Y$. The converse of Theorem 7 is given by

Theorem 8. *For every group* Y' *of ideals in* F *such that*

$$P(F, \mathfrak{t}\mathfrak{u}) \subset Y' \subset I(F, \mathfrak{t})$$

for some $\mathfrak{t}$ *and* $\mathfrak{u}$, *there exists a unique abelian extension* M *of* F *such that* $Y' \cap I(F, \mathfrak{t}\mathfrak{d})$ *is the kernel of the map*

$$I(F, \mathfrak{t}\mathfrak{d}) \ni \mathfrak{a} \mapsto [\mathfrak{a}, M/F] \in G(M/F),$$

where $\mathfrak{d}$ *is the relative discriminant of* M *over* F.

One can actually show that $Y' \subset I(F, \mathfrak{t}\mathfrak{d})$. We call this M *the class field over* F *corresponding to* Y'.

Coming back to K, $\tau_1, \ldots, \tau_g$ and K' of §6, let us take the smallest Galois extension S of Q containing K, and denote by G the Galois group of S over Q. Let H be the subgroup of G corresponding to K. Extend each τ_ν to an element of G, and denote it again by τ_ν. Put $T = \bigcup_{\nu=1}^{g} H\tau_\nu$, and

$$H' = \{\gamma \in G \mid T\gamma = T\}.$$

Then from our definition of K' (see Th. 6), we observe that K' is the subfield of S corresponding to H'. Since $H'T^{-1} = T^{-1}$, we can find elements $\sigma_1, \ldots, \sigma_h$ of G so that

$$\bigcup_{\nu=1}^{g} \tau_\nu^{-1} H = T^{-1} = \bigcup_{\lambda=1}^{h} H'\sigma_\lambda. \tag{7.2}$$

Counting the number of elements, we see that $[K' : Q] = 2h$. Moreover, for every element a (resp. ideal $\mathfrak{C}$) in K',

$$\prod_{\lambda=1}^{h} a^{\sigma_\lambda} \qquad (\text{resp. } \prod_{\lambda=1}^{h} \mathfrak{C}^{\sigma_\lambda})$$

is an element (resp. ideal) in K. This follows easily from (7.2). Now let I' be the group of all ideals $\mathfrak{C}$ in K' such that

$$\prod_{\lambda=1}^{h} \mathfrak{C}^{\sigma_\lambda} = (\beta), \quad N(\mathfrak{C}) = \beta\beta^\rho$$

with an element β of K. It can easily be seen that I' contains P(K', (1)). Now Th. 6 is refined in the following way:

<u>Theorem 6'</u>. *K' (φ(z)) is exactly the class field over K' corresponding to I'.*

Furthermore, we have an analogue of the relation (3.6). To describe it, let us denote by $A(\mathfrak{a})$ the abelian variety isomorphic to C^g/L with L defined by (6.8) for an ideal $\mathfrak{a}$ in K. Take a field k of definition for $A(\mathfrak{a})$ containing $K'(\varphi(z))$. Let σ be an isomorphism of k into C such that $\sigma = [\mathfrak{p}, K'(\varphi(z))/K']$ on $K'(\varphi(z))$ for a prime ideal $\mathfrak{p}$ in K'. Then we have

(7.3) $A(\mathfrak{a})^\sigma$ <u>is isomorphic to</u> $A(\mathfrak{a}\mathfrak{b}^{-1})$, <u>where</u> $\mathfrak{b} = \prod_{\lambda=1}^{h} \mathfrak{p}^{\sigma_\lambda}$

Further we can obtain ramified abelian extensions of K' by means of the points of finite order on A.

Let us briefly indicate how Th. 6 and (7.3) can be proved. First let us derive Th. 6' from (7.3). Let $A = A(\mathfrak{a})$ and k be as above, and τ an isomorphism of k into C. To simplify the matter, let us assume that (iii) of Th. 5 is true for the present A even if we disregard the polarization; namely assume

(7.4) A <u>is isomorphic to</u> A^τ <u>if and only if</u> $\varphi(z) = \varphi(z)^\tau$.

(This is true if $g = 1$, but not necessarily so if $g > 1$.) Now we observe that $A(\mathfrak{a})$ and $A(\mathfrak{c})$ are isomorphic if and only if $\mathfrak{a}$ and $\mathfrak{c}$ belong to the same ideal class. Therefore, the notation being as in (7.3), we see that $A(\mathfrak{a})^\sigma$ is isomorphic to $A(\mathfrak{a})$ if and only if $\mathfrak{b}$ is a principal ideal in K. Combining this fact with (7.4), we conclude that a prime ideal $\mathfrak{p}$ in K' decomposes completely in $K'(\varphi(z))$ if and only if $\prod_{\lambda=1}^{h} \mathfrak{p}^{\sigma_\lambda}$ is a principal ideal in K. This is almost the desired result, but not quite. We could not obtain the condition about $N(\mathfrak{p})$, since we disregarded the polarization. A careful analysis of polarization leads to Th. 6'.

To prove (7.3), we have to introduce the notion of reduction of an algebraic variety modulo a prime ideal. Let V be a variety in a projective space, defined over an algebraic number field k. Let $\mathfrak{P}$ be a prime ideal in k, and $k(\mathfrak{P})$ the residue field modulo $\mathfrak{P}$. We consider the set $\mathfrak{f}$ of all homogeneous polynomials vanishing on V, whose coefficients are $\mathfrak{P}$-integers. For each $f \in \mathfrak{f}$, we consider $f \bmod \mathfrak{P}$, which is a polynomial with coefficients in $k(\mathfrak{P})$. Then we define the reduction of V modulo $\mathfrak{P}$, denoted by $V[\mathfrak{P}]$, to be the set of all common zeros of the polynomials $f \bmod \mathfrak{P}$ for all $f \in \mathfrak{f}$. If V is an abelian variety defined over k, then one can show that $V[\mathfrak{P}]$ is an abelian variety defined over $k(\mathfrak{P})$ for all except a finite number of $\mathfrak{P}$. For such a $\mathfrak{P}$, reduction mod $\mathfrak{P}$ of every element of End(V) is well defined, and gives an element of End $(V[\mathfrak{P}])$.

We apply these facts to the above $A(\mathfrak{a})$. It is not difficult to obtain $A(\mathfrak{a})$ defined over an algebraic number field k. We assume that k contains $K'(\varphi(z))$ and is Galois over K'. By the principle (6.2), we can find n linearly independent linear invariant differential forms $\omega_1, \ldots, \omega_g$ on A, rational over k, so that

$$\text{(7.5)} \qquad \theta(a)^* \omega_\nu = a^{\tau_\nu} \omega_\nu \qquad (a \in \mathfrak{o}\,;\ \nu = 1, \ldots, g).$$

Let us assume, for the sake of simplicity, that K is normal over Q, $K = K'$, and the class number of K is one, though Th. 6' becomes somewhat trivial under the last condition. By (7.2), we can put $\sigma_\lambda = \tau_\lambda^{-1}$. Let $\mathfrak{P}$ be a prime ideal in K of absolute degree one, and let $\mathfrak{b} = \prod_{\lambda=1}^{g} \mathfrak{P}^{\sigma_\lambda}$. Take a prime ideal $\mathfrak{p}$ in k which divides $\mathfrak{P}$, and consider reduction modulo $\mathfrak{p}$. Indicate the reduced objects by putting tildes. From (7.5) we obtain

$$\widetilde{\theta(b)}^* \tilde{\omega}_\nu = 0 \qquad (\nu = 1, \ldots, g) \tag{7.6}$$

if $\mathfrak{b} = (b)$ with an integer b in K. Let x be a generic point of $\tilde{A}$ over $\tilde{k}$. Then the relation (7.6) shows that every derivation of $\tilde{k}(x)$ vanishes on $\tilde{k}(\widetilde{\theta(b)}x)$, hence

$$\tilde{k}(\widetilde{\theta(b)}x) \subset \tilde{k}(x^p),$$

where p is the rational prime divisible by $\mathfrak{P}$. Since

$$N(\mathfrak{b}) = p^g = [\tilde{k}(x) : \tilde{k}(x^p)],$$

we obtain

$$\tilde{k}(\widetilde{\theta(b)}x) = \tilde{k}(x^p). \tag{7.7}$$

On the other hand, if σ is the Frobenius substitution for $\mathfrak{p}$ over K, then A^σ mod $\mathfrak{p}$ can be identified with $\tilde{A}^p$. Therefore (7.7) shows that, if $A = A(\mathfrak{a})$,

(7.8) $\quad A(\mathfrak{a})^\sigma$ mod $\mathfrak{p}$ <u>is isomorphic to</u> $A(b^{-1}\mathfrak{a})$ mod $\mathfrak{p}$.

Then it is not difficult to lift the isomorphism to that of $A(\mathfrak{a})^{\sigma}$ to $A(\mathfrak{b}^{-1}\mathfrak{a})$, hence (7.3).

8. A further method of constructing class fields

As I mentioned in §3, there are some other ways of generalizing Theorem 2. For example this can be done by considering special values of automorphic functions with respect to a discrete subgroup of $SL_2(\mathbf{R})$ obtained from a quaternion algebra.

A quaternion algebra over a field F is, by definition, an algebra B over F such that $B \otimes_F \overline{F}$ is isomorphic to $M_2(\overline{F})$, where $\overline{F}$ denotes the algebraic closure of F. For our purpose we take F to be a totally real algebraic number field of finite degree. Then one can prove that

$$(8.1) \qquad B_{\mathbf{R}} = B \otimes_{\mathbf{Q}} \mathbf{R} \cong \underbrace{M_2(\mathbf{R}) \oplus \cdots \oplus M_2(\mathbf{R})}_{r} \oplus \underbrace{D \oplus \cdots \oplus D}_{g-r},$$

where D is the division ring of real Hamilton quaternions, $g = [F : \mathbf{Q}]$, and r is an integer such that $0 \leqq r \leqq g$. We assume that $r > 0$, and regard B as a subset of $B_{\mathbf{R}}$. How many such B do there exist? I shall answer this question afterward.

For $a \in B$, let $a_1, \ldots, a_r$ be the projections of a to the first r factors $M_2(R)$. We denote by B^+ the set of all a in B such that $\det(a_\nu) > 0$ for $\nu = 1, \ldots, r$. Then every element a of B^+ acts on the product $\mathfrak{H}^r$ of r copies of the upper half plane $\mathfrak{H}$, the action of each a_ν on $\mathfrak{H}$ being defined as in §2. If we denote by $N_{B/F}(a)$ the reduced norm of a to F, then B^+ is the set of all a such that $N_{B/F}(a)$ is totally positive.

Observe that B is of dimension 4g over Q. By an <u>order</u> in B, we understand a subring $\mathfrak{O}$ of B, containing Z, which is a free Z-module of rank 4g. An order is called <u>maximal</u>, if it is not contained properly in another order. There are infinitely many maximal orders in B. We fix a maximal order $\mathfrak{O}$ in B, and put

$$\Gamma = \{\gamma \in \mathfrak{O} \cap B^+ \mid \gamma\mathfrak{O} = \mathfrak{O}\}.$$

Further, for every integral ideal $\mathfrak{t}$ in F, put

$$\Gamma(\mathfrak{t}) = \{\gamma \in \Gamma \mid \gamma - 1 \in \mathfrak{t}\mathfrak{O}\}.$$

Then Γ and $\Gamma(\mathfrak{t})$, as subgroups of B^+, act on $\mathfrak{H}^r$. One can show that $\mathfrak{H}^r/\Gamma(\mathfrak{t})$ is compact if and only if B has no zero-divisor other than 0. For example, if $B = M_2(Q)$, we can set $\mathfrak{O} = M_2(Z)$, hence $\Gamma = SL_2(Z)$, and $\mathfrak{H}/\Gamma$ is not compact.

Now let us assume $r = 1$. Then $\mathfrak{H}/\Gamma(\mathfrak{t})$ is compact unless $B = M_2(Q)$. Therefore, as was discussed in §2, the quotient $\mathfrak{H}/\Gamma(\mathfrak{t})$

(or its compactification when $B = M_2(Q)$) is a compact Riemann surface, and an automorphic function with respect to $\Gamma(\mathfrak{T})$ is a meromorphic function on $\mathfrak{H}$ invariant under $\Gamma(\mathfrak{T})$ (satisfying an additional condition when $B = M_2(Q)$).

Remark. In §2 we considered only discrete subgroups of $SL_2(R)$. The action of an element α of $\Gamma(\mathfrak{T})$ on $\mathfrak{H}$ is that of α_1, i.e., the projection of α to the first factor $M_2(R)$ of B_R. The element α_1 may not be contained in $SL_2(R)$. But this does not produce any difficulty, since we only have to consider

$$\Gamma' = \{\det(\alpha_1)^{-1/2} \cdot \alpha_1 \mid \alpha \in \Gamma(\mathfrak{T})\}$$

in place of $\Gamma(\mathfrak{T})$.

The group of the above type was first introduced by Poincaré about 80 years ago in the case $F = Q$, and later Fricke considered the general case. They discussed ternary quadratic forms instead of quaternion algebras.

We have to define "special points" on $\mathfrak{H}$ relative to Γ, analogous to ω_1/ω_2 of Th. 2, where we shall examine the values of automorphic functions. For this purpose, we notice:

Lemma. Let M be a totally imaginary quadratic extension of F which is isomorphic to a quadratic subfield of B over F. Then the following assertions hold.

(1) *If f is an F-linear isomorphism of M into B, then* $f(M) - \{0\}$ *is contained in* B^{+}, *and every element of* $f(M) - F$ *has exactly one fixed point on* $\mathfrak{H}$ *which is common to all elements of* $f(M) - F$.

(2) *If* $\mathfrak{r}_M$ *denotes the ring of integers in M, then there exists an F-linear isomorphism f of M into B such that* $f(\mathfrak{r}_M) \subset \mathfrak{O}$.

The first assertion is quite easy to prove, but the second needs a somewhat deep fact of arithmetic of quaternion algebras.

We are going to take the fixed point of $f(M) - F$ as our "special point". One can of course ask a question: What kind of M can be embedded in B? Leaving this question aside for a while, we are now ready to state the main result:

Theorem 9. There exists a couple (V, φ) *formed by a projective non-singular curve V and a holomorphic mapping* φ *of* $\mathfrak{H}$ *into V with the following properties:*

(i) φ *gives a biregular isomorphism of* $\mathfrak{H}/\Gamma(\mathfrak{t})$ *into V.* (φ *is surjective unless* $B = M_2(Q)$.)

(ii) *V is defined over the class field k over F corresponding to* $P(F, \mathfrak{t}\mathfrak{u}_0)$, *where* $\mathfrak{u}_0$ *is the product of all archimedean primes of F.* (For notation, see §7.)

(iii) *Let M and f be as in (2) of the above Lemma, and z the fixed point of the elements of* $f(M) - F$ *on* $\mathfrak{H}$. *Then the composite of* $k(\varphi(z))$ *and M is exactly the class field over M corresponding to* $P(K, \mathfrak{t})$.

Thus φ and z correspond to j and ω_1/ω_2 of Th. 2. We all such a couple (V, φ) a *canonical model* for $\mathfrak{H}/\Gamma(\mathfrak{T})$. If $V, \varphi)$ and (V', φ') are two canonical models for the same $\mathfrak{H}/\Gamma(\mathfrak{T})$, ıen we can show the existence of a biregular isomorphism ξ of V onto $'$, rational over k, such that $\varphi' = \xi \circ \varphi$. In this sense, (V, φ) ı uniquely determined for $\mathfrak{H}/\Gamma(\mathfrak{T})$. It may be worth noting that the ıaximal abelian extension of M can thus be generated, over the maxi-ıal abelian extension of F, by special values of some specific automorphic ınctions.

To answer the questions about B and M, let $F_{\mathfrak{p}}$ denote the ompletion of F with respect to an archimedean or a non-archimedean rime $\mathfrak{p}$ of F. Put $B_{\mathfrak{p}} = B \otimes_F F_{\mathfrak{p}}$. Let P_B be the set of all $\mathfrak{p}$ uch that $B_{\mathfrak{p}}$ is a division algebra. Then we have the following as-ertions:

(8.2) *P_B is a finite set with an even number of primes.*

(8.3) *For any finite set P with an even number of archimedean r non-archimedean primes of F, there exists a quaternion algebra over F, unique up to F-linear isomorphism, such that $P = P_B$.*

(8.4) *A quadratic extension M of F can be F-linearly embedded ı B if and only if $M \otimes_F F_{\mathfrak{p}}$ is a field for every $\mathfrak{p} \in P_B$.*

These results are special cases of Hasses's theorems on simple lgebras over algebraic number fields. Observe that $g - r$ factors of 3.1) correspond to the archimedean primes of P_B.

The reciprocity law for the extension $M.k(\varphi(z))$ over M can e described explicitly in terms of the special points $\varphi(z)$. For implicity, let us consider only the case where the class number of

F in the narrow sense is one, i.e., $P(F, \mathfrak{n}_0) = I(F, (1))$. For every prime ideal $\mathfrak{p}$ in F, let $\mathfrak{r}_{\mathfrak{p}}$ be the ring of $\mathfrak{p}$-integers in $F_{\mathfrak{p}}$, and let $\mathfrak{O}_{\mathfrak{p}} = \mathfrak{r}_{\mathfrak{p}} \mathfrak{O}$. Then $\mathfrak{O}_{\mathfrak{p}}$ is a maximal order in $B_{\mathfrak{p}}$. Let $U(\mathfrak{t})$ denote the group of all elements α of B^+ such that α is a unit of $\mathfrak{O}_{\mathfrak{p}}$ for all $\mathfrak{p}$ dividing $\mathfrak{t}$, and let $U_0(\mathfrak{t})$ be the subgroup of $U(\mathfrak{t})$ consisting of all α such that $\alpha \equiv 1 \bmod \mathfrak{t}\mathfrak{O}_{\mathfrak{p}}$ for all $\mathfrak{p}$ dividing $\mathfrak{t}$. It can easily be shown that $U(\mathfrak{t})/U_0(\mathfrak{t})$ is isomorphic to $(\mathfrak{O}/\mathfrak{t}\mathfrak{O})^{\times}$ (see Notation). For every $\alpha \in U(\mathfrak{t})$, put

$$\sigma(\alpha) = [(N_{B/F}(\alpha)), k/F].$$

Now we have

Theorem 10. There exists a system of biregular isomorphisms $R(\alpha)$ of V to $V^{\sigma(\alpha)}$, defined for each $\alpha \in U(\mathfrak{t})$ and rational over k, with the following properties:

(i) $R(\alpha)^{\sigma(\beta)} \circ R(\beta) = R(\alpha\beta)$.

(ii) $R(\alpha) = R(\beta)$ if $\alpha^{-1}\beta \in U_0(\mathfrak{t})$.

(iii) $R(\gamma)[\varphi(z)] = \varphi(\gamma(z))$ for every $z \in \mathfrak{H}$ if $\gamma \in \Gamma(1)$.

(iv) Let M, f, and z be as in (iii) of Th. 9 (still under the condition $f(\mathfrak{r}_M) \subset \mathfrak{O}$). Suppose that f is normalized in the sense defined below. Let $\mathfrak{b}$ be an ideal in M prime to $\mathfrak{t}$, and let

$$\tau = [\mathfrak{b}, M.k(\varphi(z))/M].$$

Then there exists an element α of $U(\mathfrak{t})$ such that $f(\mathfrak{b})\mathfrak{O} = \alpha\mathfrak{O}$.

<u>(Such an α is not unique.) With such an element α, one has</u>

$$\varphi(z)^{\tau} = R(\alpha)[\varphi(\alpha^{-1}(z))].$$

Here we say that f is <u>normalized</u> if

$$\frac{d}{dw}[f(a)(w)]_{w=z} = \bar{a}/a \qquad (0 \neq a \in M).$$

If we define $\bar{f}$ by $\bar{f}(a) = f(\bar{a})$ for $a \in M$, then we see that either f or $\bar{f}$ is normalized.

It should be observed that (iv) of Th. 10 is a generalization of (3.7). Further, if $B = M_2(Q)$, $\mathfrak{O} = M_2(Z)$, and $\mathfrak{t} = (N)$ with a positive integer N, the function field of V is exactly the field $Q(j, f^N_{ab})$ considered in §3. Therefore the first three properties of $R(\alpha)$ in Th. 10 may be regarded as a generalization of Th. 4.

<u>Example</u>. Let us consider the case $F = Q(\zeta + \zeta^{-1})$ with $\zeta = e^{2\pi i/d}$ for d = 7, 9, or 11. By (8.3), there exists a unique quaternion algebra B over F, for which P_B consists of all but one archimedean primes of F, the exception being the archimedean prime of F corresponding to the identity map of F. The present F has the class number one in the narrow sense. Moreover, all the maximal orders in B are conjugate to each other under the inner automorphisms of B defined by the elements of B^+. Take a maximal order $\mathfrak{O}$ in B, and

define $\Gamma = \Gamma(1)$ as above. Then one can prove that $\mathfrak{H}/\Gamma$ is of genus 0, and Γ modulo its center is generated by three elements $\gamma_2, \gamma_3, \gamma_d$ of order 2, 3, d, respectively, satisfying $\gamma_2\gamma_3\gamma_d = 1$. (If $d = 7$, the measure given by (3.4) is $1/42$, which is the smallest value of (3.4) for all possible Γ.) These three elements have unique fixed points on $\mathfrak{H}$, one for each, which we denote by w_2, w_3, w_d. Then there exists a unique meromorphic function φ on $\mathfrak{H}$ such that $C(\varphi)$ is the field of all automorphic functions on $\mathfrak{H}$ with respect to Γ, and $\varphi(w_2) = 1$, $\varphi(w_3) = 0$, $\varphi(w_d) = \infty$. If we denote by V the complex projective line, then φ gives a biregular isomorphism of $\mathfrak{H}/\Gamma$ onto V. Now we can prove that this (V, φ) <u>is a canonical model</u> for $\mathfrak{H}/\Gamma$. By (8.4), for <u>every</u> totally imaginary quadratic extension M of F, there exists a F-linear isomorphism f of M into B such that $f(\mathfrak{r}_M) \subset \mathfrak{O}$. Let $\{z_1, \ldots, z_q\}$ be a set of representatives for the Γ-equivalence classes of the fixed points of $f(M) - F$ for all such f. Then q is exactly the class number of M, and from (iii) of Th. 9 and (iv) of Th. 10, we obtain:

(8.5) <u>The values</u> $\varphi(z_1), \ldots, \varphi(z_q)$ <u>form a complete set of conjugates of</u> $\varphi(z_1)$ <u>over</u> F, <u>and for each</u> i, $M(\varphi(z_i))$ <u>is the maximal unramified abelian extension of</u> M.

Thus φ has a strong resemblance to the classical modular function j.

Unfortunately, the proofs of Theorems 9 and 10 are long and very complicated. Therefore I have to content myself with a rough sketch of the main ideas. We take a totally imaginary quadratic extension K of F and consider (A, c, θ) of type (K, Φ) in the sense of §6 with a representation Φ of K such that

$$\Phi \sim \Sigma_{\nu=1}^{r}(\tau_\nu + \rho\tau_\nu) + \Sigma_{\nu=r+1}^{g} 2\tau_\nu ,$$

where τ_ν and ρ are as in §6. Then it can be shown that the (A, c, θ) of this type are parameterized by the point on $\mathfrak{H}^r$, and there is a discontinuous group Γ' acting on $\mathfrak{H}^r$ such that $\mathfrak{H}^r/\Gamma'$ is in one-to-one correspondence with all the isomorphism classes of such (A, c, θ). Taking K suitably, we can identify Γ' with the above Γ (in the case r = 1). For this family of (A, c, θ), we can find a model (V', φ') of $\mathfrak{H}/\Gamma'$ with the properties analogous to those of Th. 5. If M and z are as in (iii) of Th. 9, the corresponding (A, c, θ) is such that $\mathrm{End}_{\mathbf{Q}}(A)$ contains an isomorphic image of $K \otimes_F M$. The coordinates of $\varphi'(z)$ generate an abelian extension of the nature described in Th. 6'. This couple (V', φ') is the first approximation to the desired (V, φ). From infinitely many such (V', φ'), depending on the choice of K, we can construct a canonical model (V, φ).

If $B = M_2(\mathbf{Q})$, there is a family of elliptic curves, for which the value of j is the modulus. For the basic field F of higher degree, there is no such <u>standard</u> family of abelian varieties, though infinitely many families of abelian varieties can be loosely associated with a given $\mathfrak{H}/\Gamma$. It is an open question whether there exists any family of geometric structures, other than the above (A, c, θ), of which our canonical model (V, φ) is a <u>natural</u> variety of moduli.

9. The Hasse zeta function of an algebraic curve

Let V be a projective non-singular curve of genus h defined over an algebraic number field k. For every prime ideal $\mathfrak{p}$ in k, let $k(\mathfrak{p})$ denote the residue field modulo $\mathfrak{p}$. Considering the equations for V modulo $\mathfrak{p}$, we obtain a curve $V[\mathfrak{p}]$ over $k(\mathfrak{p})$ (see §7). It can be shown that $V[\mathfrak{p}]$ is non-singular, and of genus h for almost all $\mathfrak{p}$. (We call such $\mathfrak{p}$ good.) Therefore one can speak of the zeta function of $V[\mathfrak{p}]$ over $k(\mathfrak{p})$ which is of the form

$$Z_{\mathfrak{p}}(u) = Z^1_{\mathfrak{p}}(u)/[(1-u)(1-N(\mathfrak{p})u)],$$

where u is an indeterminate, and $Z^1_{\mathfrak{p}}(u)$ is a polynomial of degree $2h$. The Hasse zeta function of V over k, denoted by $Z(s;V/k)$ with complex variable s, is now defined by

$$Z(s;V/k) = \prod_{\mathfrak{p}} Z^1_{\mathfrak{p}}(N(\mathfrak{p})^{-s})^{-1},$$

the product being taken over all "good" $\mathfrak{p}$. (It is important to consider also "bad" $\mathfrak{p}$, which we shall not discuss here.) Now one can make the following

<u>Conjecture</u>. $Z(s;V/k)$ <u>can be continued to the whole</u> s-<u>plane and satisfies a functional equation</u>.

The purpose of this section is to verify this conjecture for the curve which is a canonical model for $\mathfrak{H}/\Gamma(\mathfrak{t})$ in the sense of §8. For the sake of simplicity, we shall consider here only the case where $\mathfrak{t} = (1)$ and the class number of F in the narrow sense is one, i.e., every ideal in F is a principal ideal generated by a totally positive element. In this case, for every right $\mathfrak{O}$-ideal $\mathfrak{a}$, there exists an element a in B^+ such that $\mathfrak{a} = a\mathfrak{O}$.

Let us introduce cusp forms and Hecke operators with respect to the group $\Gamma = \Gamma(1)$. For every $\xi = \begin{pmatrix} a & b \\ c & d \end{pmatrix} \in GL_2(R)$ with $\det(\xi) > 0$, put

$$j(\xi, z) = \det(\xi)^{1/2} / (cz + d).$$

Let m be a positive integer. By a _cusp form of weight_ m _with respect to_ Γ, we understand a holomorphic function $f(z)$ on $\mathfrak{H}$ satisfying the following two conditions:

(i) $f(\gamma(z))j(\gamma, z)^m = f(z)$ _for all_ $\gamma \in \Gamma$.

(ii) _If_ s _is a cusp of_ Γ, _and_ ρ, q _are as in_ §2, _then_ $f(\rho(z))j(\rho, z)^m$ _is holomorphic in_ $q = 2\pi iz$, _and vanishes at_ $q = 0$.

The latter condition is necessary only when $B = M_2(Q)$, since Γ has no cusps otherwise. All such functions form a vector space of finite dimension over C, which we write $S_m(\Gamma)$. If $m = 2$, the map

$$S_2(\Gamma) \ni f(z) \mapsto f(z)dz$$

gives an isomorphism of $S_2(\Gamma)$ onto the space of differential forms of the first kind on $\mathfrak{H}/\Gamma$, hence the dimension of $S_2(\Gamma)$ is equal to the genus of $\mathfrak{H}/\Gamma$.

For every $\alpha \in B^+$, we note that the double coset $\Gamma\alpha\Gamma$ can be decomposed into a finite number of one sided cosets:

$$\Gamma\alpha\Gamma = \bigcup_{i=1}^{e} \alpha_i \Gamma \qquad \text{(disjoint).}$$

Then for $f \in S_m(\Gamma)$, we define $(\Gamma\alpha\Gamma)_m f$ by

$$(\Gamma\alpha\Gamma)_m f = \Sigma_{i=1}^{e} f(\alpha_i^{-1}(z)) j(\alpha_i^{-1}, z)^m .$$

In this way we obtain a linear transformation $(\Gamma\alpha\Gamma)_m$ on $S_m(\Gamma)$, which is of course independent of the choice of the representatives α_i. For every integral ideal $\mathfrak{a}$ in F, let $T(\mathfrak{a})_m$ denote the sum of all distinct $(\Gamma\alpha\Gamma)_m$ such that $\alpha \in \mathfrak{O}$ and $(N_{B/F}(\alpha)) = \mathfrak{a}$. Then we define a Dirichlet series $D_m(s)$, whose coefficients are linear endomorphisms of $S_m(\Gamma)$, by

$$D_m(s) = \Sigma\, T(\mathfrak{a})_m N(\mathfrak{a})^{-s},$$

where $\mathfrak{a}$ runs over all integral ideals in F. It can be shown that $D_m(s)$ converges for sufficiently large $\mathrm{Re}(s)$ and has an Euler product:

$$D_m(s) = \prod [1 - T(\mathfrak{p})_m N(\mathfrak{p})^{-s}]^{-1}$$

$$\times \prod [1 - T(\mathfrak{p})_m N(\mathfrak{p})^{-s} + N(\mathfrak{p})^{1-2s}]^{-1},$$

where the first product is taken over all the prime ideals $\mathfrak{p}$ in P_B

(see (8.2-4)), and the second over the remaining primes in F. Moreover, $D_m(s)$ can be holomorphically continued to the whole s-plane, and satisfies a functional equation:

$$R_m(s) = G_m(s)D_m(s) = T_o \cdot R_m(2-s),$$

$$T_o = (-1)^{m/2} \prod_{\mathfrak{p} \in P_B^o} T(\mathfrak{p}),$$

$$G_m(s) = (2\pi)^{-gs} \prod_{\mathfrak{p} \in P_B^o} N(\mathfrak{p})^{s/2}\, \Gamma(s - 1 + m/2)\Gamma(s)^{g-1},$$

where P_B^o means the set of non-archimedean $\mathfrak{p}$ in P_B. The last few Γ stand for the usual gamma function.

Now we have

Theorem 11. Let (V, φ) be a canonical model for $\mathfrak{H}/\Gamma$. Then, for almost all prime ideals $\mathfrak{p}$ in F, the zeta function of V mod $\mathfrak{p}$, over the residue field mod $\mathfrak{p}$, coincides with

$$\det[1 - T(\mathfrak{p})_2 u + N(\mathfrak{p})u^2] / [(1-u)(1 - N(\mathfrak{p})u)]$$

(i.e., $Z^1_{\mathfrak{p}}(u)$ is the Euler $\mathfrak{p}$-factor of $D_m(s)$ with $m = 2$ and $N(\mathfrak{p})^{-s} = u$). Thus $Z(s; V/F)$ coincides with $\det(D_2(s))$ up to a finite number of $\mathfrak{p}$-factors.

The proof of this theorem is roughly as follows. For every $\alpha \in B^+$, define a subset $X(\Gamma\alpha\Gamma)$ of $V \times V$ by

$$X(\Gamma\alpha\Gamma) = \{\varphi(z) \times \varphi(\alpha(z)) \mid z \in \mathfrak{H} \cup (\text{cusps})\}.$$

We see that $X(\Gamma\alpha\Gamma)$ is an algebraic correspondence, which depends only on $\Gamma\alpha\Gamma$ and not on the choice of α. We observe that $T(\mathfrak{p})_2 = (\Gamma\alpha\Gamma)_2$ with any element α of $\mathfrak{O} \cap B^+$ such that $\mathfrak{p} = (N_{B/F}(\alpha))$. For such an α, we write $X(\Gamma\alpha\Gamma)$ as $X_{\mathfrak{p}}$.

Let M, f, and z be as in (iii) of Th. 9 and (iv) of Th. 10. Take M so that $\mathfrak{p}$ decomposes into two prime ideals $\mathfrak{q}$ and $\mathfrak{q}'$ in M. We can find an element β of $\mathfrak{O} \cap B^+$ so that $f(\mathfrak{q})\mathfrak{O} = \beta\mathfrak{O}$. Then $\mathfrak{p} = N_{M/F}(\mathfrak{q}) = (N_{B/F}(\beta))$, hence $(\Gamma\beta\Gamma)_2 = T(\mathfrak{p})_2$ by the above remark. Therefore $\varphi(\beta^{-1}(z)) \times \varphi(z) \in X_{\mathfrak{p}}$. Now let $\tau = [\mathfrak{q}, M(\varphi(z))/M]$. By (iv) of Th. 10, we have $\varphi(z)^{\tau} = \varphi(\beta^{-1}(z))$, hence, putting $y = \varphi(z)$, we obtain $y^{\tau} \times y \in X_{\mathfrak{p}}$. Consider now reduction modulo a prime divisor of a suitably large field, which divides $\mathfrak{p}$, and denote the reduced object by tilde. Then $\widetilde{y^{\tau}} = \tilde{y}^{N(\mathfrak{p})}$, hence

$$\tilde{y}^{N(\mathfrak{p})} \times \tilde{y} \subset \tilde{X}_{\mathfrak{p}} .$$

Let Π denote the Frobenius correspondence on $\tilde{V} \times \tilde{V}$, i.e., the locus of $x \times x^{N(\mathfrak{p})}$ for $x \in \tilde{V}$. The above discussion shows that $\tilde{X}_{\mathfrak{p}}$ has infinitely many points in common with ${}^t\Pi$, the transpose of Π. Since ${}^t\Pi$ is irreducible, we have ${}^t\Pi \subset \tilde{X}_{\mathfrak{p}}$. From our definition of $X_{\mathfrak{p}}$, it can easily be seen that ${}^tX_{\mathfrak{p}} = X_{\mathfrak{p}}$. Therefore $\Pi + {}^t\Pi \subset \tilde{X}_{\mathfrak{p}}$. Again from the definition of $X_{\mathfrak{p}}$, we see that $\tilde{X}_{\mathfrak{p}} \cdot (x \times \tilde{V})$ consists of $N(\mathfrak{p}) + 1$ points for a generic x on $\tilde{V}$.

Therefore the equality

$$\widetilde{X}_{\mathfrak{p}} = \Pi + {}^t\Pi \tag{9.1}$$

should hold.

Let Δ denote the diagonal of $V \times V$. Then

$$N(\mathfrak{p}).\Delta = \Pi \circ {}^t\Pi . \tag{9.2}$$

The relations (9.1) and (9.2) show that any symmetric function of Π and ${}^t\Pi$ can be obtained from a correspondence of V by reduction modulo $\mathfrak{p}$. In particular, for every positive integer m, we have

$$\Pi^m + {}^t\Pi^m = P_m(\widetilde{X}_{\mathfrak{p}}, N(\mathfrak{p})\Delta)$$

for a polynomial P_m, which is determined by

$$-\frac{d}{du}\log(1 - xu + yu^2) = \sum_{m=1}^{\infty} P_m(x, y)u^{m-1} \tag{9.3}$$

with indeterminates x, y, and u.

Let $I[Y]$ denote the number of fixed points of a correspondence Y, i.e., the intersection number of Y with Δ. If $Z_{\mathfrak{p}}(u)$ denotes the zeta function of $V \bmod \mathfrak{p}$, then

$$\begin{aligned}\frac{d}{du}\log Z_{\mathfrak{p}}(u) &= \Sigma_{m=1}^{\infty} I[\Pi^m]u^{m-1}\\ &= (1/2)\, \Sigma_{m=1}^{\infty} I[\Pi^m + {}^t\Pi^m]u^{m-1}\\ &= (1/2)\, \Sigma_{m=1}^{\infty} I[P_m(\widetilde{X}_{\mathfrak{p}}, N(\mathfrak{p})\Delta)]u^{m-1}.\end{aligned}$$

But if Y is a correspondence on $V \times V$, we have $I[Y] = I[\tilde{Y}]$, and by the Lefschetz fixed point theorem,

$$I[Y] = \Sigma_{i=0}^{2} \operatorname{tr}(Y \mid H^i(V)),$$

where $H^i(V)$ denotes the usual real i-th cohomology group of V. For $Y = X(\Gamma\alpha\Gamma)$, it is easy to see that

$$\operatorname{tr}(Y \mid H^1(V)) = 2. \operatorname{Re}[\operatorname{tr}(\Gamma\alpha\Gamma)_2],$$

since $S_2(\Gamma)$ is isomorphic to the space of differential forms of the first kind on $V = \mathfrak{H}/\Gamma$. Also it is obvious that

$\operatorname{tr}(Y \mid H^0(V)) = \operatorname{tr}(Y \mid H^2(V)) =$ the number of right cosets in $\Gamma\alpha\Gamma$.

One can further show that $\operatorname{tr}(\Gamma\alpha\Gamma)_2$ is real. Therefore we have

$$\begin{aligned} &\frac{d}{du} \log Z_{\mathfrak{p}}(u) \\ &= \Sigma_{m=1}^{\infty}[1 + N(\mathfrak{p})^m]u^{m-1} + \Sigma_{m=1}^{\infty} \operatorname{tr}[P_m(T(\mathfrak{p})_2, N(\mathfrak{p}))]u^{m-1}. \end{aligned}$$

In view of (9.3), we obtain

$$\frac{d}{du} \log Z^1_{\mathfrak{p}}(u) = \frac{d}{du} \log \left\{ \det[1 - T(\mathfrak{p})_2 u + N(\mathfrak{p})u^2]^{-1} \right\},$$

hence our theorem.

10. Infinite Galois extensions with ℓ-adic representations.

So far we have been interested only in the construction of abelian extensions. Now we are going to show that the above canonical model for $\mathfrak{H}/\Gamma(\mathfrak{t})$ can be employed to obtain meaningful non-abelian extensions of a number field, with some pleasant features.

Let us call $(V, \varphi, R(a))$ of Th. 10 a <u>canonical system of level</u> $\mathfrak{t}$. For every integral ideal $\mathfrak{a}$ in F, we fix a canonical system of level $\mathfrak{a}$, and denote it by $(V_{\mathfrak{a}}, \varphi_{\mathfrak{a}}, R_{\mathfrak{a}}(a))$. (We are still assuming that the class number of F is one in the narrow sense, though our discussion can actually be done without this assumption.) Let $k_{\mathfrak{a}}$ denote the class field over F corresponding to $P(F, \mathfrak{a}\mathfrak{n}_0)$, where $\mathfrak{n}_0$ is the product of all archimedean primes of F (see §7). As is stated in Th. 9, 10, $V_{\mathfrak{a}}$ and $R_{\mathfrak{a}}(a)$ are defined over $k_{\mathfrak{a}}$. If $\mathfrak{a} \subset \mathfrak{b}$, we can obtain a rational map

$$T_{\mathfrak{b},\mathfrak{a}} : V_{\mathfrak{a}} \rightarrow V_{\mathfrak{b}}, \tag{10.1}$$

defined over $k_{\mathfrak{a}}$, such that

$$T_{\mathfrak{b},\mathfrak{a}} \circ \varphi_{\mathfrak{a}} = \varphi_{\mathfrak{b}}, \quad T_{\mathfrak{b},\mathfrak{a}}^{\sigma_{\mathfrak{a}}(a)} \circ R_{\mathfrak{a}}(a) = R_{\mathfrak{b}}(a) \circ T_{\mathfrak{b},\mathfrak{a}},$$

for every $a \in U(\mathfrak{a})$, where

$$\sigma_{\mathfrak{a}}(a) = [(N_{B/F}(a)), k_{\mathfrak{a}}/F].$$

When $\mathfrak{a} = \mathfrak{b}$, this means the uniqueness of canonical system of level $\mathfrak{a}$. In the general case, the map (10.1) defines a Galois covering. If we consider the curves over the universal domain C, then the Galois group is isomorphic to $\Gamma(\mathfrak{b})/\Gamma(\mathfrak{a})E_{\mathfrak{b}}$, where $E_{\mathfrak{b}}$ denotes the group of all units e of F such that $e \equiv 1 \mod \mathfrak{b}$. If we take $k_{\mathfrak{a}}$ as the field of definition, we obtain:

<u>Proposition 1. Let</u> y' <u>be an arbitrary point of</u> $V_{\mathfrak{a}}$ <u>and</u> $y = T_{\mathfrak{b},\mathfrak{a}}(y')$. ($y'$ <u>may be generic or algebraic.</u>) <u>Then</u> $k_{\mathfrak{a}}(y')$ <u>depends only on</u> y, <u>and is a finite Galois extension of</u> $k_{\mathfrak{b}}(y)$. <u>For every</u> $\tau \in G(k_{\mathfrak{a}}(y')/k_{\mathfrak{b}}(y))$, <u>there exists an element</u> β <u>of</u> $U(\mathfrak{a}) \cap U_0(\mathfrak{b})$ <u>such that</u> $y'^{\tau} = R_{\mathfrak{a}}(\beta)(y')$, <u>and</u> $\tau = \sigma_{\mathfrak{a}}(\beta)$ <u>on</u> $k_{\mathfrak{a}}$.

<u>Proposition 2. Take a point</u> z <u>of</u> $\mathfrak{H}$ <u>so that</u> $\varphi_{\mathfrak{a}}(z) = y'$, <u>and let</u> $\Gamma^z = \{\gamma \in \Gamma(\mathfrak{b}) \mid \gamma(z) = z\}$. <u>Then the following statements hold:</u>

(i) $\Gamma(\mathfrak{a}).\Gamma^z = \{\gamma \in \Gamma(\mathfrak{b}) \mid R_{\mathfrak{a}}(\gamma)(y') = y'\}$.

(ii) <u>From the correspondence</u> $\tau \mapsto \beta$ <u>described in Prop. 1, one obtains an isomorphism of</u> $G(k_{\mathfrak{a}}(y')/k_{\mathfrak{b}}(y))$ <u>into</u>

$$[U(\mathfrak{a}) \cap U_0(\mathfrak{b})]/U_0(\mathfrak{a})\Gamma^z.$$

(iii) <u>If</u> y <u>is generic on</u> $V_{\mathfrak{b}}$ <u>over</u> $k_{\mathfrak{b}}$, <u>this isomorphism is surjective, and</u> $\Gamma^z = E_{\mathfrak{b}}$.

Let us now fix an integral ideal $\mathfrak{t}$ in F and an arbitrary point y on $V_{\mathfrak{t}}$. Take a point $y_{\mathfrak{a}}$ on $V_{\mathfrak{a}}$ for each $\mathfrak{a} \subset \mathfrak{t}$ so that $T_{\mathfrak{b},\mathfrak{a}}(y_{\mathfrak{a}}) = y_{\mathfrak{b}}$ if $\mathfrak{a} \subset \mathfrak{b}$, and $y_{\mathfrak{t}} = y$. (For example, choose a point z on $\mathfrak{H}$ so that $\varphi_{\mathfrak{t}}(z) = y$, and put $y_{\mathfrak{a}} = \varphi_{\mathfrak{a}}(z)$.) Let $\mathcal{K}_y$ denote the composite of the $k_{\mathfrak{a}}(y_{\mathfrak{a}})$ for all $\mathfrak{a}$. By Prop. 1, $\mathcal{K}_y$ is a Galois extension of $k_{\mathfrak{t}}(y)$. Our purpose is to investigate the Galois group of $\mathcal{K}_y$ over $k_{\mathfrak{t}}(y)$ and the behavior of the Frobenius automorphisms (when y is algebraic) with respect to certain representations of the Galois group.

For every prime ideal $\mathfrak{l}$ in F, let $F_{\mathfrak{l}}$, $\mathfrak{r}_{\mathfrak{l}}$, $B_{\mathfrak{l}}$, and $\mathfrak{O}_{\mathfrak{l}}$ be the localizations as defined in §8. Form the product $\mathfrak{U}$ of the groups $\mathfrak{O}_{\mathfrak{l}}^{\times}$ (see Notation) for all $\mathfrak{l}$, with the usual product topology. Let $\mathfrak{U}_{\mathfrak{t}}$ denote the subgroup of $\mathfrak{U}$ consisting of the elements $(u_{\mathfrak{l}})$ with $u_{\mathfrak{l}} \in \mathfrak{O}_{\mathfrak{l}}^{\times}$ such that $u_{\mathfrak{l}} \equiv 1 \bmod \mathfrak{t}\mathfrak{O}_{\mathfrak{l}}$ for every $\mathfrak{l}$.

For simplicity, we assume

$$(10.2) \qquad \{\gamma \in \Gamma(\mathfrak{t}) \mid \gamma(z) = z\} = E_{\mathfrak{t}}$$

for a point z on $\mathfrak{H}$ such that $\varphi_{\mathfrak{t}}(z) = y$. This is satisfied for all except a finite number of points on $V_{\mathfrak{t}}$. Under the assumption, we are going to define an injection

$$J : G(\mathcal{K}_y / k_{\mathfrak{t}}(y)) \rightarrow \mathfrak{U}_{\mathfrak{t}} / \bar{E}_{\mathfrak{t}} ,$$

where $\bar{E}_{\mathfrak{t}}$ is the closure of $E_{\mathfrak{t}}$ in $\mathfrak{U}$. Let $\tau \in G(\mathcal{K}_y / k_{\mathfrak{t}}(y))$. For every $\mathfrak{a} \subset \mathfrak{t}$, we find, by Prop. 1, an element $\xi_{\mathfrak{a}} = \xi(\mathfrak{a})$ of $U(\mathfrak{a}) \cap U_0(\mathfrak{t})$, so that $y_{\mathfrak{a}}^{\tau} = R_{\mathfrak{a}}(\xi_{\mathfrak{a}})(y_{\mathfrak{a}})$. It is not difficult to choose the elements $\xi_{\mathfrak{a}}$ so that

$$\xi_{\mathfrak{a}} \in \mathfrak{O}, \quad \xi_{\mathfrak{a}} \equiv \xi_{\mathfrak{b}} \bmod \mathfrak{b} \text{ if } \mathfrak{a} \subset \mathfrak{b}.$$

For each prime ideal $\mathfrak{l}$, the sequence $\{\xi(\mathfrak{l}^n)\}_{n=1,2,\ldots}$ converges to an element $u_{\mathfrak{l}}$ of $\mathfrak{O}_{\mathfrak{l}}^{\times}$. We define $J(\tau)$ to be the element of $\mathfrak{U}_{\tau}/\bar{E}_{\tau}$ represented by $(u_{\mathfrak{l}})$. We can verify that J is actually a continuous injection. J depends on the choice of the sequence of points $\{y_{\alpha}\}$. But it is unique up to inner automorphisms of $\mathfrak{U}_{\tau}/\bar{E}_{\tau}$. Further we have:

<u>Proposition</u> 3. <u>If</u> y <u>is generic on</u> V_{τ} <u>over</u> k_{τ}, J <u>is surjective</u>.

Let us now discuss the points y whose coordinates are algebraic numbers; our main interest is of course in this case. First we consider the point fixed by an imaginary quadratic subfield of B. Let M, f, and z be as in (iii) of Th. 9, and $\mathfrak{r}_M$ the ring of integers in M. Then $f(\mathfrak{r}_M) \subset \mathfrak{O}$. Put

$$\mathcal{R} = \prod_{\mathfrak{l}} f(\mathfrak{r}_M)_{\mathfrak{l}}^{\times}$$

and view $\mathcal{R}$ as a subgroup of $\mathfrak{U}$. Then, from (iv) of Th. 10, we obtain

<u>Proposition</u> 4. <u>If</u> $y = \varphi_{\tau}(z)$ <u>and</u> $M \subset k_{\tau}(y)$, <u>then the image of</u> J <u>is contained in</u> $(\mathcal{R} \cap \mathfrak{U}_{\tau})/\bar{E}_{\tau}$.

If M is not contained in $k_{\tau}(y)$, we have to consider a larger group which contains $\mathcal{R} \cap \mathfrak{U}_{\tau}$ as a subgroup of index 2, but we shall not go into details.

Next let us consider the case of an arbitrary algebraic point y of V_τ. Take any central simple algebra A over Q, and consider a representation

$$\rho : B^\times \to A^\times$$

satisfying the following two conditions:

(1) ρ <u>is rational over</u> Q, $B^\times$ <u>and</u> $A^\times$ <u>being considered as algebraic groups over</u> Q.

(2) $\rho(a) = N_{F/Q}(a)^m$ <u>for every</u> $a \in F^\times$ <u>with an integer</u> m, <u>independent of</u> a.

For every rational prime ℓ, let Q_ℓ denote the field of ℓ-adic integers, and let $B_\ell = B \otimes_Q Q_\ell$, $A_\ell = A \otimes_Q Q_\ell$. From ρ one can naturally obtain a representation

$$\rho_\ell : B_\ell^\times \to A_\ell^\times .$$

Let $\mathfrak{U}^\ell$ denote the product of the groups $\mathfrak{o}_{\mathfrak{l}}^\times$ for all prime factors $\mathfrak{l}$ of ℓ, and E_τ^ℓ the closure of E_τ in $\mathfrak{U}^\ell$. Put $\mathfrak{U}_\tau^\ell = \mathfrak{U}^\ell \cap \mathfrak{U}_\tau$. Combining J with a natural homomorphism of $\mathfrak{U}_\tau/\overline{E}_\tau$ to $\mathfrak{U}_\tau^\ell/E_\tau^\ell$, we obtain a homomorphism

$$J_\ell : G(\mathfrak{K}_y/k_\tau(y)) \to \mathfrak{U}_\tau^\ell/E_\tau^\ell .$$

Now we make the following assumption:

(3) $\rho(e) = 1$ <u>for all</u> $e \in E_{\mathfrak{r}}$.

In view of (2), this is automatically satisfied if either m is even, or the units of $E_{\mathfrak{r}}$ are all totally positive. Under the assumption, observe that $\rho_\ell \circ J_\ell$ is meaningful, and defines a representation of $G(\mathfrak{K}_y / k_{\mathfrak{r}}(y))$ into $A_\ell^\times$. Now it is important to investigate the behavior of the Frobenius automorphisms of $\mathfrak{K}_y$ with respect to the "ℓ-adic representations" $\rho_\ell \circ J_\ell$. First we notice:

<u>Theorem</u> 12. <u>There exists an integral ideal</u> $\mathfrak{m}$ <u>in</u> $k_{\mathfrak{r}}(y)$ <u>such that every prime ideal of</u> $k_{\mathfrak{r}}(y)$, <u>prime to</u> $\ell\mathfrak{m}$, <u>is unramified in the subfield of</u> $\mathfrak{K}_y$ <u>which corresponds to the kernel of</u> $\rho_\ell \circ J_\ell$.

Therefore, if $\mathfrak{p}$ is such a prime ideal of $k_{\mathfrak{r}}(y)$, $\mathfrak{P}$ is a prime divisor of $\mathfrak{K}_y$ dividing $\mathfrak{p}$, and $\sigma_{\mathfrak{P}}$ is a Frobenius automorphism of $\mathfrak{K}_y$ over $k_{\mathfrak{r}}(y)$ for $\mathfrak{P}$, then the conjugacy class of $\rho_\ell \circ J_\ell(\sigma_{\mathfrak{P}})$ is determined only by $\mathfrak{p}$. In this setting, we have the following theorem, which may be regarded as the main result of this section:

<u>Theorem</u> 13. <u>The roots of the principal polynomial of</u> $\rho_\ell \circ J_\ell(\sigma_{\mathfrak{P}})$ <u>over</u> Q_ℓ <u>are algebraic numbers of absolute value</u> $N(\mathfrak{p})^{m/2}$.

This fact leads us to a temptation of making the following conjectures.

<u>Conjecture</u> I. <u>The principal polynomial of</u> $\rho_\ell \circ J_\ell(\sigma_{\mathfrak{P}})$ <u>has rational coefficients, and is independent of</u> ℓ.

Suppose this is true, and let $f_{\mathfrak{p}}$ denote the principal polynomial of $\rho_{\ell} \circ J_{\ell}(\sigma_{\mathfrak{p}})$. Then one can define a <u>zeta function</u> $\zeta(s; \mathfrak{K}_y; \rho)$ of $\mathfrak{K}_y$ <u>associated with the representation</u> ρ <u>by</u>

$$\zeta(s; \mathfrak{K}_y; \rho) = \prod_{\mathfrak{p}} N(\mathfrak{p})^{ns} . f_{\mathfrak{p}}(N(\mathfrak{p})^s)^{-1},$$

where the product is extended over all prime ideals $\mathfrak{p}$ in $k_{\mathfrak{r}}(y)$ prime to $\mathfrak{m}$, and n is the degree of $f_{\mathfrak{p}}$.

<u>Conjecture II.</u> $\zeta(s; \mathfrak{K}_y; \rho)$ <u>can be analytically continued to the whole</u> s<u>-plane and satisfies a functional equation.</u>

Further, as for the nature of $\mathfrak{K}_y$, one may make

<u>Conjecture III.</u> <u>The image of</u> J <u>is an open subgroup of</u> $\mathfrak{U}_{\mathfrak{r}}/\bar{E}_{\mathfrak{r}}$, <u>unless</u> $y = \varphi_{\mathfrak{r}}(z)$ <u>with a point</u> z <u>such that</u>

$$\{\alpha \in B^+ \mid \alpha(z) = z\} = f(M - \{0\})$$

<u>for some totally imaginary quadratic extension</u> M <u>of</u> F <u>and an</u> F<u>-linear isomorphism</u> f <u>of</u> M <u>into</u> B.

It should be remarked that the above representation ρ is analogous to the Grössen-character in Hecke's sense, or more precisely, to the Grössen-character of type (A_0) in the sense of Taniyama-Weil. If $y = \varphi_{\mathfrak{r}}(z)$ with a point z fixed by $f(M - \{0\})$ as excluded in Conjecture III, and if $M \subset k_{\mathfrak{r}}(y)$, then we can show that $\zeta(s; \mathfrak{K}_y; \rho)$ is a product of several zeta functions of $k_{\mathfrak{r}}(y)$ with Grössen-characters. One may also notice that $\mathfrak{U}_{\mathfrak{r}}/\bar{E}_{\mathfrak{r}}$ is analogous to the idèle class group of a number field modulo the connected component. Further, if $B = M_2(Q)$, $\mathfrak{O} = M_2(Z)$, $\mathfrak{r} = NZ$ with a positive integer N, then $\Gamma(\mathfrak{r})$ is the principal congruence subgroup of $SL_2(Z)$ of level N, and the choice of a point y on $V_{\mathfrak{r}}$ is almost equivalent to the choice of an elliptic curve. Thus, if ρ is the identity mapping of $B^{\times} = GL_2(Q)$ to itself, $\zeta(s, \mathcal{R}_y, \rho)$ is the Hasse zeta function of the curve corresponding to y. A similar fact holds also when $F = Q$ and B is a division algebra. It is an open question whether such an interpretation exists for $\zeta(s, \mathcal{R}_y, \rho)$ in the case where F is of degree > 1. In this connection, it should be mentioned that the extension $\mathcal{R}_y$ is rather different from the extension obtained from the points of finite order on an abelian variety, if $[F : Q] > 1$.

11. Further generalization and concluding remarks

We have obtained two different types of results: one is represented by Th. 5, Th. 6, and Th. 6'; the other by Th. 9 and Th. 10. They are however two special cases of a more general theorem. To see this, let us introduce discontinuous groups which include $Sp(n, Z)$ and $\Gamma(\mathfrak{r})$ as special cases.

Let F, B and r be as in §8. Since $B \otimes_F \overline{F} = M_2(F)$ for the algebraic closure $\overline{F}$ of F, we can regard the elements of B as matrices of degree 2 with entries in $\overline{F}$. For every $\alpha = \begin{pmatrix} a & b \\ c & d \end{pmatrix} \in M_2(\overline{F})$, put

$$\alpha^* = \begin{pmatrix} d & -b \\ -c & a \end{pmatrix} = \delta . {}^t\alpha \delta^{-1}, \quad \delta = \begin{pmatrix} 0 & 1 \\ -1 & 0 \end{pmatrix}.$$

One can show that $\alpha^* \in B$ if $\alpha \in B$, and $\alpha \mapsto \alpha^*$ is an involution of B in the sense of §6, which is not necessarily positive. For an element $U = (\alpha_{ij}) \in M_n(B)$ with $\alpha_{ij} \in B$, put ${}^tU^* = {}^t(\alpha_{ij}^*)$. Then $U \mapsto {}^tU^*$ defines an involution of $M_n(B)$. This involution can be R-linearly extended to $M_n(B_R)$, where $B_R = B \otimes_Q R$. Define a Lie group G by

$$G = \left\{ U \in M_n(B_R) \mid {}^tU^*U = 1_n \right\}.$$

In view of (8.1), we have

$$M_n(B_R) \cong \underbrace{M_{2n}(R) \oplus \dots \oplus M_{2n}(R)}_{r} \oplus \underbrace{M_n(D) \oplus \dots \oplus M_n(D)}_{g-r}.$$

According to this direct sum decomposition, G can be decomposed into a direct product:

$$G = G_1 \times \dots \times G_r \times G_{r+1} \times \dots \times G_g .$$

One can easily show that

$$G_\nu \cong Sp(n, R) \cong \left\{ X \in SL_{2n}(R) \mid {}^tXJ_nX = J_n \right\}, \quad J_n = \begin{pmatrix} 0 & -1_n \\ 1_n & 0 \end{pmatrix},$$

if $\nu \leqq r$,

$$G_\nu \cong \left\{ X \in GL_n(D) \mid {}^t\bar{X}X = 1_n \right\} \qquad \text{if } \nu > r,$$

where the bar means the quaternion conjugate in D. Therefore, $G_{r+1}, \ldots, G_g$ are compact. Since $Sp(n, R)$ acts on the Siegel space $\mathfrak{H}_n$, we can let every element U of G act on the product $\mathfrak{H}_n^r$ of r copies of $\mathfrak{H}_n$, the action of U on the ν-th factor $\mathfrak{H}_n$ being that of the projection of U to G_ν. As in §8, take a maximal order $\mathfrak{O}$ in B, and put

$$\Gamma = G \cap M_n(\mathfrak{O}).$$

In this way we obtain a discontinuous group Γ acting on $\mathfrak{H}_n^r$. If $B = M_2(Q)$ and $\mathfrak{O} = M_2(Z)$, Γ is $Sp(n, Z)$. If $n = 1$, the present Γ is a subgroup of finite index of the group Γ considered in §8. The quotient $\mathfrak{H}_n^r/\Gamma$ is compact in the following two cases: (i) $r < g$; (ii) $r = g$, $n = 1$, and B is a division algebra. The group Γ was introduced by Siegel in his paper on symplectic geometry (under the restriction $r = 1$). Now, for this quotient $\mathfrak{H}_n^r/\Gamma$, we can construct a couple (V, φ) with the properties analogous to those in Th. 9, thus unifying the above mentioned two types of results. A part of the results of §10 can also be extended to such a general case.

One can investigate automorphic functions with respect to a more general type of group. Namely one takes a semi-simple algebraic group $\mathfrak{G}$ defined over Q and consider a Lie group $\mathfrak{G}_R$ consisting of the points with coefficients in R. Suppose that the quotient of $\mathfrak{G}_R$

by a maximal compact subgroup is a bounded symmetric domain. Then one can speak of (meromorphic) automorphic functions and forms on this domain with respect to the group $\Gamma = \mathfrak{G}_Z$ formed by integral points on $\mathfrak{G}$. There are many interesting arithmetical problems in this field, which I dare not enumerate here. But I should at least mention that almost all important questions are related to automorphic forms and zeta functions explicitly or implicitly, on which I have talked only in § 9.

12. Bibliography

Among a vast multitude of literature, I shall try to list standard reference books from the view point of accessibility and (probable) comprehensibility, along with a few recent papers relevant to the topics discussed in these lectures.

The reader with a standard knowledge of algebraic groups or Lie theory may find the following volume useful:

[1] Algebraic groups and discontinuous subgroups, Proceedings of Symposia in Pure Mathematics, vol. 9, Amer. Math. Soc. 1966.

This contains many interesting surveys of recent investigations, most of which have abundant references. For those who are more interested in the classical modular functions or modular forms, many important papers in

[2] E. Hecke, Mathematische Werke, Göttingen, 1959

will serve as standard references. A more systematic and somewhat easier treatment is presented in

[3] M. Eichler, Einführung in die Theorie der algebraischen Zahlen und Funktionen, Birkhäuser, 1963 (the English revised version is available).

At the end of each chapter of this book, there are plenty of references. As a textbook on the classical theory of elliptic functions, the following may be recommended:

[4] C. Jordan, Cours d'analyse de l'école polytéchnique, Paris, vol. II, Ch. VII.

As for the theory of complex multiplication of elliptic functions, I pick here only two, old and (relatively) new:

[5] H. Weber, Lehrbuch der Algebra III, 2nd ed., 1908,

[6] M. Deuring, Die klassenkörper der komplexen Multiplication, Enzyclopädie d. math. Wiss. neue Aufl. Bd. I_2, Heft 10_{II}, Stuttgart, 1958.

The fundamental material of abelian varieties is presented by

[7] A. Weil, Variétés abéliennes et courbes algébriques, Hermann, Paris, 1948.

[8] S. Lang, Abelian varieties, Interscience, New York, 1959.

The analytic theory of theta functions and abelian varieties is systematically treated in

[9] A. Weil, Introduction a l'étude des variétés kählériennes, Hermann, Paris, 1958,

[10] C. L. Siegel, Analytic functions of several complex variables, lecture notes, Institute for Advanced Study, 1948, reprinted with corrections, 1962.

The latter will serve also as an introduction to the theory of automorphic functions of several variables. On this topic and other related subjects, one can not miss

[11] C. L. Siegel, Gesammelte Abhandlungen, 3 vol., Springer, 1966.

Especially for Siegel modular functions, the standard knowledge can be obtained from

[12] H. Maass, Lectures on Siegel's modular functions, Tata Institute, 1954-55,

[13] Séminaire H. Cartan, 1957/58, Fonctions automorphes.

A detailed account of the results discussed in §7 on the nature of the number field $K'(\varphi(z))$ etc. is presented in

[14] G. Shimura and Y. Taniyama, Complex multiplication of abelian varieties and its applications to number theory, Publ. Math. Soc. Japan, No. 6, Tokyo, 1961.

A recent volume

[15] A. Weil, Basic number theory, Springer, 1967

contains a modern treatment of class field theory, as well as the structure theorems of simple algebras over number fields, which generalize (8.2-4). The latter subject, in a concise style, can be found in

[16] M. Deuring, Algebren, Ergebn, der Math., Berlin, 1935.

As for the general theory of arithmetically defined discontinuous groups, I mention here only three papers:

[17] A. Borel and Harish-Chandra, Arithmetic subgroups of algebraic groups, Ann. of Math. 75 (1962), 485-535.

[18] G. D. Mostow and T. Tamagawa, On the compactness of arithmetically defined homogeneous spaces, Ann. of Math. 76 (1962), 446-463.

[19] W. L. Baily and A. Borel, Compactification of arithmetic quotients of bounded symmetric domains, Ann. of Math. 84 (1966), 442-528.

The compactness criterion, which generalizes that for $\mathfrak{H}_n^r/\Gamma$ of §11, is given in [17] and [18]. The last paper [19] proves the existence of a Zariski open subset of a projective variety isomorphic to a given quotient like $\mathfrak{H}_n^r/\Gamma$ in general. For these topics, see also the articles in [1].

Theorems 5, 9, 10, 11 and their generalizations are proved in my papers:

[20] G. Shimura, Moduli and fibre systems of abelian varieties, Ann. of Math. 83 (1966), 294-338.

[21] G. Shimura, Construction of class fields and zeta functions of algebraic curves, Ann. of Math. 85 (1967), 58-159.

[22] G. Shimura, Algebraic number fields and symplectic discontinuous groups, Ann. of Math. 86 (1967), 503-592.

The last section of [22] is a weaker version of the results stated in §10, for which a full account will be discussed in a forthcoming paper. Some basic concepts of ℓ-adic representations can be found in

[23] Y. Taniyama, L-functions of number fields and zeta functions of abelian varieties, J. Math. Soc. Japan, 9 (1957), 330-366.

This verifies also the Hasse conjecture for abelian varieties with sufficiently many complex multiplications. For this topic, see also [14], and lecture notes by J.-P. Serre to be published soon. The connection of the Hasse zeta function of an elliptic curve with the Diophantine problems on the curve is discussed in a survey article (of course with numerous references)

[24] J. W. S. Cassels, Diophantine equations with special reference to elliptic curves, J. London Math. Soc., 41 (1966), 193-291.